MUĞLA ÜNİVERSİTESİ

TÜBİTAK

UNESCO

# Biosaline Agriculture and Salinity Tolerance in Plants

Edited by M. Öztürk, Yoav Waisel, M. Ajmal Khan and Güven Görk

**Birkhäuser Verlag**
**Basel • Boston • Berlin**

Münir Öztürk
Ege University
Center for Environmental Studies
Science Faculty Building A-Blok
35100 Bornova
Izmir-Turkey

Yoav Waisel
Department of Plant Sciences
The George S. Wise Faculty of Life Sciences
Tel Aviv University
Tel Aviv
Israel

M. Ajmal Khan
Department of Botany
University of Karachi
Karachi-75270
Pakistan

Güven Görk
Department of Biology
Faculty of Science and Arts
Mugla University
Mugla-Turkey



A CIP catalogue record for this book is available from the Library of Congress, Washington D.C., USA

**Bibliographic information published by Die Deutsche Bibliothek**
Die Deutsche Bibliothek lists this publication in the Deutsche Nationalbibliografie; detailed bibliographic data is available in the Internet at <http://dnb.ddb.de>.

ISBN 10: 3-7643-7609-0 Birkhäuser Verlag, Basel – Boston – Berlin
ISBN 13: 978-3-7643-7609-3



Printed on acid-free paper produced from chlorine-free pulp. TCF ∞

Cover design: Micha Lotrovsky, CH-4106 Therwil

Cover illustration: From the top: Naturally growing field of *Salicornia europaea* (M. Öztürk); example of Biolog GP2 Metabolic fingerprints of taproot rhizosphere bacterial communities (Y. Waisel, see page 6); a general view of halophytes from an inland saline habitat (M. Öztürk); *Centaurea spinosa* from a coastal habiat (M. Öztürk).
Typesetting: PTP-Berlin Protago-TEX-Production GmbH, Germany

ISBN 10: 3-7643-7609-0
ISBN 13: 978-3-7643-7609-3

e-ISBN 10: 3-7643-7610-4
e-ISBN 13: 978-3-7643-7610-9

9 8 7 6 5 4 3 2 1

www.birkhauser.ch

# Contents

# List of contributors

Chedly Abdelly, Laboratoire d'Adaptation des Plantes aux Stress Abiotiques, INRST, BP 95, Hammam-Lif, 2050, Tunisia; e-mail: chedly.abdelly@inrst.rnrt.tn

Ahmet Aksoy, University of Erciyes, Science and Art Faculty, Dept. of Biology, 38039 Kayseri, Turkey; e-mail: aksoy@erciyes.edu.tr

Muhammad Ashraf, Department of Botany, University of Agriculture, Faisalabad (38040), Pakistan; e-mail: shrafm@fsd.paknet.com.pk

Muhammad Y. Ashraf, Stress Physiology and Plant Biochemistry Lab, Salinity and Environmental Management Division, Nuclear Institute for Agriculture and Biology, P.O. Box 18, Jhang Road, Faisalabad, Pakistan; e-mail: niabmyashraf@hotmail.com

Abdallah Atia, Laboratoire d'Adaptation des Plantes aux Stress Abiotiques, INRST, BP 95, Hammam-Lif, 2050, Tunisia

Zouhaier Barhoumi, Adaptation des Plantes aux Stress Abiotiques, INRST, BP 95, Hammam-Lif 2050, Tunisia

Nader Ben Amor, Laboratoire d'Adaptation des Plantes aux Stress Abiotiques, INRST, BP 95, Hammam-Lif, 2050, Tunisia

Karim Ben Hamed, Laboratoire d'Adaptation des Plantes aux Stress Abiotiques, INRST, BP 95, Hammam-Lif, 2050, Tunisia

Clanton C. Black, Department of Biochemistry and Molecular Biology, University of Georgia, Life Sciences Building, Anthens, GA 30602-7229, USA

Ahmed Debez, Laboratoire d'Adaptation des Plantes aux Stress Abiotiques, INRST, BP 95, Hammam-Lif, 2050, Tunisia

Hassan M. El Shaer, Animal and Poultry Production Division, Desert Research Center, P.O. Box 11753, Mataria, Cairo, Egypt; e-mail: drc_elshaer@hotmail.com

Shahina A. Ghazanfar, Royal Botanic Gardens, Kew, Richmond, Surrey, TW9 3AB, UK

Tahar Ghnaya, Laboratoire d'Adaptation des Plantes aux Stress Abiotiques, INRST, BP 95, Hammam-Lif, 2050, Tunisia

Lilya G. Gismatullina, Department of Desert Ecology and Water Resources Research, Samarkand Branch of Academy of Sciences, 3 Timur Malik Str., Samarkand 703000, Uzbekistan

Çiğdem Görk, Mugla University, Biology Department, Mugla, Turkey

Güven Görk, Mugla University, Biology Department, Mugla, Turkey

Claude Grignon, Biochimie et Physiologie Moléculaire des Plantes, Agro-M INRA, 34060 Montpellier, France

Aykut Güvensen, Ege University, Biology Department, 35100 Bornova-Izmir, Turkey
Ergin Hamzaoğlu, University of Erciyes, Science and Art Faculty of Yozgat, Department of Biology, Yozgat, Turkey; e-mail: ehamzaoglu@erciyes.edu.tr
Bernhard Huchzermeyer, Institut für Botanik, Universität Hannover Herrenhäuserstr. 2, 30419 Hannover, Germany
Faqir Hussain, Stress Physiology and Plant Biochemistry Lab, Salinity and Environmental Management Division, Nuclear Institute for Agriculture and Biology, P.O. Box 18, Jhang Road, Faisalabad, Pakistan
Muhammad M. Iqabal, Stress Physiology and Plant Biochemistry Lab, Salinity and Environmental Management Division, Nuclear Institute for Agriculture and Biology, P.O. Box 18, Jhang Road, Faisalabad, Pakistan
H. Hüsnü Kayikçioğlu, Ege University Faculty of Agriculture, Department of Soil Science, Bornova-Izmir, 35100 Turkey; e-mail: husnu.kayikcioglu@ege.edu.tr
M. Ajmal Khan, Department of Botany, University of Karachi, Karachi 75270, Pakistan
Halil Kirnak, University of Harran, Agriculture Faculty, Irrigation Department, 63200 Sanliurfa, Turkey; e-mail: hkirnak@yahoo.com
Fatih Konukcu, Trakya University, Tekirdag Agricultural Faculty, Irrigation and Drainage Department, 59030 Tekirdag, Turkey;
e-mail: fatih.konukcu@tu.tzf.edu.tr
Ashwani Kumar, Biotechnology Laboratory, Department of Botany, University of Rajasthan, Jaipur-Rajasthan, 302004, India
Riadh Ksouri, Adaptation des Plantes aux Stress Abiotiques, INRST, BP 95, Hammam-Lif 2050, Tunisia
Nahla Laabidi, Laboratoire d'Adaptation des Plantes aux Stress Abiotiques, INRST, BP 95, Hammam-Lif, 2050, Tunisia
Victoria V. Li, Department of Desert Ecology and Water Resources Research, Samarkand Branch of Academy of Sciences, 3 Timur Malik Str., Samarkand 703000, Uzbekistan
Weiqiang Li, Center for Agricultural Resources Research, Institute of Genetics and Developmental Biology, Chinese Academy of Sciences, 286 Huaizhong Raod, Shijiazhuang, Hebei 050021, P.R. China
Xiaojing Liu, Center for Agricultural Resources Research, Institute of Genetics and Developmental Biology, Chinese Academy of Sciences, 286 Huaizhong Raod, Shijiazhuang, Hebei 050021, P.R. China; e-mail: xjliu@ms.sjziam.ac.cn
Wolfgang Maibaum, Agricultural Advisory Department, K+S Kali GmbH, Bertha-von Suttner-Strasse 7, 34131 Kassel, Germany
Gulnara K. Mardonova, Department of Desert Ecology and Water Resources Research, Samarkand Branch of Academy of Sciences, 3 Timur Malik Str., Samarkand 703000, Uzbekistan
Dorsaf Messedi, Laboratoire d'Adaptation des Plantes aux Stress Abiotiques, INRST, BP 95, Hammam-Lif, 2050, Tunisia
Ghulam Murtaza, Institute of Soil and Environmental Sciences, University of Agriculture, Faisalabad 38040, Pakistan
Karl-Hermann Neumann, Institut für Pflanzenernährung, Abteilung Gewebekultur, IFZ, Heinrich-Buff-Ring 26-32, 35392 Giessen Germany

Maya Ofek, Department of Plant Sciences, Tel Aviv University, Ramat Aviv 69978, Tel Aviv, Israel

Bülent Okur, Ege University Faculty of Agriculture, Department of Soil Science, Bornova-Izmir, 35100 Turkey; e-mail: bulent.okur@ege.edu.tr

Jim D. Oster, Department of Environmental Sciences, Unversity of California, Riverside, CA 92521, USA

Zeineb Ouerghi, Laboratoire d'Adaptation des Plantes aux Stress Abiotiques, INRST, BP 95, Hammam-Lif, 2050, Tunisia

Münir Öztürk, Botany Department, Science Faculty, Ege University, 35100 Bornova-Izmir, Turkey; e-mail: munirozturk@hotmail.com

Manzoor Qadir, International Center for Agricultural Research in the Dry Areas, P.O. Box 5466, Aleppo, Syria; e-mail: M.Qadir@cgiar.org

Hailong Qiao, Center for Agricultural Resources Research, Institute of Genetics and Developmental Biology, Chinese Academy of Sciences, 286 Huaizhong Raod, Shijiazhuang, Hebei 050021, P.R. China

Annamaria Ranieri, Dipartimento Chimica e Biotecnologie Agrarie, Università degli studi di Pisa, Pisa, Italy

Marcus Ross, Agricultural Advisory Department, K+S Kali GmbH, Bertha-von Suttner-Strasse 7, 34131 Kassel, Germany

Silke Ruppel, Institute of Vegetable and Ornamental Crops Grossbeeren/Erfurt e.V., Theodor-Echtermeyer-Weg 1, 14979 Grossbeeren, Germany

Dhouha Saadaoui, Laboratoire d'Adaptation des Plantes aux Stress Abiotiques, INRST, BP 95, Hammam-Lif, 2050, Tunisia

Ghulam Sarwar, Stress Physiology and Plant Biochemistry Lab, Salinity and Environmental Management Division, Nuclear Institute for Agriculture and Biology, P.O. Box 18, Jhang Road, Faisalabad, Pakistan

Arnould Savour, Unité Physiologie Cellulaire et Moléculaire des Plantes, CNRS-UMR 7632, 4 Place Jussieu, Paris, France

Sven Schubert, Institute of Plant Nutrition, Justus Liebig University, Heinrich-Buff-Ring 26-32, 35392 Giessen, Germany

Vijendra P.S. Shekhawat, Biotechnology Laboratory, Department of Botany, University of Rajasthan, Jaipur-Rajasthan, 302004, India;
e-mail: vijendrapss@yahoo.com

Elena V. Shuiskay, Department of Desert Ecology and Water Resources Research, Samarkand Branch of Academy of Sciences, 3 Timur Malik Str., Samarkand 703000, Uzbekistan

Ines Slama, Laboratoire d'Adaptation des Plantes aux Stress Abiotiques, INRST, BP 95, Hammam-Lif, 2050, Tunisia

Abderrazzak Smaoui, Adaptation des Plantes aux Stress Abiotiques, INRST, BP 95, Hammam-Lif 2050, Tunisia

Abdelaziz Soltani, Laboratoire d'Adaptation des Plantes aux Stress Abiotiques, INRST, BP 95, Hammam-Lif, 2050, Tunisia

Wael Taamalli, Laboratoire de Caractérisation de Qualité de l'Huile d'Olive, INRST, BP 95, Hammam-Lif, 2050 Tunisia

Toshiaki Tadano, Department of Applied Biology and Chemistry, Faculty of Applied Bio-Science, Tokyo University of Agriculture, 1-1-1 Sakuraoka, Setagaya-ku, Tokyo, 156-8502, Japan

Ons Talbi, Adaptation des Plantes aux Stress Abiotiques, INRST, BP 95, Hammam-Lif 2050, Tunisia

Kristina N. Toderich, Department of Desert Ecology and Water Resources Research, Samarkand Branch of Academy of Sciences, 3 Timur Malik Str., Samarkand 703000, Uzbekistan; e-mail: ktoderich@yahoo.com

Riaz A. Wahed, Stress Physiology and Plant Biochemistry Lab, Salinity and Environmental Management Division, Nuclear Institute for Agriculture and Biology, P.O. Box 18, Jhang Road, Faisalabad, Pakistan

Yoav Waisel, Department of Plant Sciences, Tel Aviv University, Ramat Aviv 69978, Tel Aviv, Israel; e-mail: waisel@post.tau.ac.il

Temur R. Yunusov, Department of Desert Ecology and Water Resources Research, Samarkand Branch of Academy of Sciences, 3 Timur Malik Str., Samarkand 703000, Uzbekistan

Mokhtar Zarrouk, Laboratoire de Caractérisation de Qualité de l'Huile d'Olive, INRST, BP 95, Hammam-Lif, 2050 Tunisia

Fethia Zribi, Adaptation des Plantes aux Stress Abiotiques, INRST, BP 95, Hammam-Lif 2050, Tunisia

# Foreword

Global demand for the precious resource of water has increased six-fold over the past century, with a three-fold increase in world population. The water crisis is one of the most critical challenges facing the world today.

Seawater is globally available in abundance, and hypersaline soils are widespread. Based on science, and with carefully established good practices, large areas of saline soils can be converted into high productivity man-made agro-ecosystems.

Substantial information has been provided by numerous scientists since the early 1960s, regarding the restoration, functioning, and development of saline ecosystems and halophytes, and international centres and societies have been established.

We have to be ready to respond when land becomes non-productive due to high salt concentrations. The availability of correct and adequate scientific knowledge is absolutely essential to develop good biosaline management practices.

UNESCO has supported a number of such activities, societies and centres, and this is well in tune with the organisation's focus on *water and associated ecosystems*.

It is now important to identify the next important milestones. A concerted international action is required in order to continue the process of advancing science-based biosaline practices, and to develop profitable models and products. It is also important to raise public awareness: some marketable products have already been developed, and provide valuable services to mankind, such as *Salicornia bigelovii*, *Salicornia europaea*, *Aster tripolium* (vegetable and salad), *Conocarpus erecta*, *Conocarpus lanciofolius* (roadside trees), and *Sesuvium portulacastrum* (to replace freshwater dependant ground cover), to name only a few.

The Arab States in the Gulf suffer greatly from a lack of freshwater availability, whereas saline groundwater and seawater occur in abundance.

It is with this in mind, that the UNESCO Office in Doha, and in agreement with the UNESCO Office in Venice, decided to support Arab experts to participate in *The International Conference on Biosaline Agriculture and Salinity Tolerance in Plants*, Mugla University, Turkey, in January 2005, as well as with this important publication.

The book has three sections: the first section deals with physiological aspects of salt tolerance. It provides data and new information regarding a number of moderate to high salinity tolerant plants species, such as *Vicia faba*, a cash crop, several grass species, as well as *Crithmum maritimum*, *Suaeda salsa*, *Salsola* spp, *Atriplex centralasiatica*, *Cakile maritima*, as well as the seawater tolerant *Sesuvium portulacastrum.*

The second section provides new information on ecological aspects, such as biological diversity conservation, management of natural plant diversity, geographical inventories of halophyte communities, and vegetation zones.

The third section on agriculture provides valuable information on the utilisation of halophytes, soil irrigation and drainage management, bio-reclamation of saline soils, and effects of salinity on crop productivity.

UNESCO congratulates the editors and authors of this book, who produced an excellent scholarly work. Improving the knowledge of the multidisciplinary audience of readers will contribute towards improvement of scientific research, education, and environmental management.

*Biosaline Agriculture and Salinity Tolerance in Plants* is another important scientific contribution towards the management of salt-affected soils, saline irrigation water, and halophytes.

BENNO BÖER
Ecological Sciences Advisor – Arab Region
*UNESCO Regional Office in the Arab States of the Gulf*
*Doha, PO Box 3945, State of Qatar*
*b.boer@unesco.org*

February 2006

# Acknowledgements

The contributions of 127 scientists from 16 countries covering both scientific as well as practical concepts regarding the agricultural production and environmental protection on salt-affected ecosystems is highly appreciated. We gratefully acknowledge the financial support given to us by the Muğla University, Turkish National Research Council, UNESCO-Doha and Toros Gübre-Muğla, Turkey, which made it possible to hold the International Conference on "Biosaline Agriculture and High Salinity Tolerance" in Muğla, Turkey. We would like to express our gratitude to the Governor of Muğla, Mayor of Muğla Dr. Osman Gürün, Rector Muğla University Prof. Dr. Şener Oktik and Vice Rector Prof. Dr. Ibrahim Yokaş, for their keen interest in the organisation of this conference. Finally, our sincerest appreciation goes to the local Organizing Committee in Muğla, in particular Dr. Çiğdem Görk and Dr. Kübra Karaosmanoğlu for their untiring efforts to assure both the scientific and social success of the conference.

Last but not the least we would like to thank the Birkhäuser Verlag, for their help and flexibility, as well as professional handling of the publishing process.

Biosaline Agriculture and Salinity Tolerance in Plants
Edited by M. Öztürk, Y. Waisel, M.A. Khan and G. Görk


# Effects of salinity on rhizosphere bacterial communities associated with different root types of *Vicia faba L.*

Maya Ofek[1], Silke Ruppel[2] and Yoav Waisel[1]

[1] *Department of Plant Sciences, Tel Aviv University, Ramat Aviv 69978, Tel Aviv, Israel*
[2] *Institute of Vegetable and Ornamental Crops Grossbeeren/Erfurt e.V., Theodor-Echtermeyer-Weg 1, 14979, Grossbeeren, Germany*

## Introduction

Soils constitute a heterogeneous and patchy environment that can be divided into innumerable microhabitats. Each of these portray different conditions of temperature, salinity, nutrient composition, abundance and availability, aeration, etc. Conceptually, the various roots of each root system have different environments to adapt to. Such adaptations are manifested by changes in root properties and activities [1].

Comparative investigations of tap and lateral roots of young faba bean plants (*Vicia faba* L.) have shown differences in nutrient uptake, water uptake and growth. Differences were also found in abscisic acid content and in the response of the roots to stimuli by applied hormones. The number of isozymes of several enzymes and their activities differed between taproots and laterals. Such differences were greatly emphasized under stress conditions of salinity, hypoxia and nutrient deprivation [2]. Moreover, it has been shown that the development of such traits is genetically controlled and that a set of genes is specific for each root type [3].

Plant root systems can be classified into different types, each having distinct inherent morphological, physiological and biochemical characteristics [4]. It is postulated that such traits determine the plants' capability to cope with their spatially and temporally heterogeneous environment. It was thus tempting to assume that the physiological differences between roots should also be expressed by changes in their respective rhizospheres at different soil horizons.

Rhizosphere colonization by microorganisms is affected by various environmental conditions [1, 5, 6]. Indeed, root exudates, sloughed-off cells and disintegrating tissues attract bacteria and are the main contributors to the enrichment of the rhizosphere microbiota [6]. Abundance and activity of microbial communities are much higher in the rhizosphere than in the bulk soil [5, 7, 8]. Investigations of rhizosphere bacterial communities were firstly based on culture dependent – but more recently on culture independent – molecular methods [9–12]. Some investigated root-bacteria interactions were found under different environmental conditions, between different roots and even between different segments along roots [13]. Nevertheless, specificity of association between different root types of individual plants and bacterial communities has hardly been tested.

In the following investigation we tackled the question whether various roots of single root systems of *Vicia faba* support different rhizosphere bacterial communities even when grown under homogeneous saline and non-saline conditions.

## Material and methods

### *Plant growth conditions*

Faba bean plants were grown in an aeroponic growth system (25°C, 12 h photoperiod, photon flux: 170 $\mu$mol m$^{-2}$ sec$^{-1}$) for 18 days. Half strength Hoagland's nutrient solution [14] was supplemented with soil bacterial extracts and 50 mM NaCl was added to some of the treatments. The solution was replaced every 3 days in order to achieve a constant and even supply of nutrients.

Soil samples (20 g samples of a sieved grumosol) were extracted in 80 mL saline solution (0.85% NaCl), shaken and centrifuged. The pellets were re-suspended in saline solution (0.85% NaCl) and aliquots were then added to the aeroponic nutrient solutions. Bacterial cell densities of the soil extracts were determined as colony forming units (CFU) ml$^{-1}$ by plating on Laurie-broth (LB) medium. Cell densities ranged between $4 \times 10^8$ and $6 \times 10^9$ CFU ml$^{-1}$.

### *Root sampling*

Roots of 18 days old plants were washed with sterilized water. Terminal 5 cm root segments of the tap and of the lateral roots were transferred into sterile saline solutions (1:10 w:v). Rhizosphere bacterial communities (RBC) were separated from the roots by vigorous vortex for 10 min. The resulting suspensions were diluted ($\times 10^3$) and immediately inoculated into Biolog microplates.

### *Biolog assay*

Biolog GP2 microplates were used for bacterial community analysis. Plates were inoculated with 150 $\mu$L RBC suspensions per well and incubated at 28°C for up to 96 h. Color development was expressed as light absorbance ($A$) at $\lambda = 595$ nm and was measured at 24 h intervals using the SpectraMAX 190 absorbance microplate reader.

Samples of the obtained RBCs were decimally diluted and plated on LB growth medium and were incubated at 28°C for 4 days. Cell densities were determined and expressed as CFU ml$^{-1}$. Distinguishable bacterial morphotypes were picked from the plates of each extract and isolated on LB medium.

*Taxonomic identification*

Taxonomic identification of the conspicuous morphotypes was done using the recommended Biolog system (GP2 and GN2). For a selected number of isolates, 16S rDNA genes were amplified by PCR using the primer set 27f and 1429r [15] (Tab. 1). Sequence results were compared to known 16S rDNA sequences available in the gene bank using the BLAST module.

*Data handling and analyses*

Biolog GP2 blanked $A_{595}$ of each plate and at each recording time was calculated by subtraction of the blank well reading from each of the carbon source wells. Average well color development (AWCD) values were calculated as: $\Sigma$ (well $A_{595}$ minus blank well $A_{595}/95$ [16]. Threshold values were set as the minimal $A_{595}$ of the wells in which color could clearly be detected. Binary transformed data was used for non-metric distance analysis (NMDA) using the STATISTICA software package (version 6.0).

Means of the ordination positions were calculated for the 1$^{st}$ and 2$^{nd}$ dimensions in NMDA and compared by two-way ANOVA ($p < 0.05$).

Diversity of C source utilization in the Biolog plates was analyzed [17]. Shannon-Weiner index of diversity (H′) was calculated as: $H' = -\Sigma p_i \ln p_i$ where $p_i$ is the proportion of color development in the $i$th well over the total color development of the wells.

## Results

Differences between the RBCs of lateral and of taproots were found, both under saline and under non-saline conditions. The root types can be distinguished by differences in the utilization patterns of different carbon source groups as exemplified in Figure 1. These differences were expressed in NMDA (Fig. 2).

The root type strongly affected the RBCs carbon source utilization diversity, expressed by Shannon-Weiner index of diversity (Tab. 2). H′ differed between lateral roots and taproots RBCs whether examined for the complete set of carbon sources or for particular C groups (Tab. 2). Polymers were an exception as H′ of this group was similar between root types and treatments throughout all measurements. The effect of salinity was minor, as compared to the effect of the root type. H′ calculated for the complete set of carbon sources differed between the RBCs of the treatments after 48 h of incubation though not after 72 h and 96 h. The group of sugars was the only carbon group in which H′ of the treatment and control RBCs differed significantly (Tab. 2).

In order to clarify a taxonomic basis for the differences in carbon source utilization patterns, isolated bacterial pure cultures were individually identified using the Biolog GN system. Results indicated higher number of genera for lateral roots RBCs as compared to those of the taproots (Tab. 3). Out of a total of eight identified genera, seven genera were found in the lateral roots RBCs, while only three genera were

**Table 1.** Oligonucleotide primer sequences and PCR conditions used in this investigation

| Target taxon | Primer sequences (forward and reverse) | Target gene | Annealing T°C |
|---|---|---|---|
| *Pseudomonas aeruginosa* | oprL(f): 5′-ATG GAA ATG CTG AAA TTC GGC-3′<br>oprL(r): 5′-CTT CTT CAG CTC GAC GCG ACG-3′ | *oprL* | 64 |
| *Herbaspirillum* | HRS (f): 5′-GCA AGA CCT CAT GCT CCT G-3′<br>HRS (r): 5′-CAC GGC TAG AGT GTG TC-3′ | *16S rDNA* | 62 |
| *Eubacteria* | 27f: 5′-AGA GTT TGA TCC TGG CTC AG-3′<br>1492r: 5′- TAC CTT GTT ACG ACT T-3′ | *16s rDNA* | 56 |
| *Eubacteria* | 519f: 5′-CAG C(A/C)G CCG CGG TAA (A/C/G/T)(A/T)C-3′<br>907r: 5′- CCG TCA ATT C(A/C)T TT(A/G) AGT T-3′ | *16S rDNA* | 53 |

**Table 2.** Diversity of carbon source utilization by taproots and by lateral roots rhizosphere bacterial communities of *Vicia faba* in Biolog GP2 microplates. Diversity was expressed as Shannon-Weiner index of diversity (H′) and was calculated after 48, 72 and 96 h of incubation of the microplates for the complete set of carbon sources and for the different carbon source groups. Means of the H′ values of three replicates for each RBC were compared by ANOVA ($p < 0.05$). Means ± standard deviations are presented ($*: P < 0.05$; $**: p < 0.01$; $***: p < 0.001$).

| | Inc. (h) | Root types | | | Treatments | | |
|---|---|---|---|---|---|---|---|
| | | Lateral roots | Taproots | | Non-saline | Saline | |
| *Complete set* | 48 | 2.74 ± 0.51 | 3.11 ± 0.17 | * | 3.11 ± 0.18 | 2.74 ± 0.51 | * |
| | 72 | 3.21 ± 0.22 | 3.47 ± 0.14 | ** | 3.39 ± 0.09 | 3.28 ± 0.31 | |
| | 96 | 3.34 ± 0.18 | 3.60 ± 0.15 | *** | 3.49 ± 0.07 | 3.45 ± 0.30 | |
| *Sugars* | 48 | 1.72 ± 0.72 | 1.84 ± 0.28 | | 2.07 ± 0.40 | 1.49 ± 0.50 | * |
| | 72 | 2.25 ± 0.34 | 2.50 ± 0.17 | * | 2.46 ± 0.16 | 2.28 ± 0.37 | * |
| | 96 | 2.41 ± 0.26 | 2.69 ± 0.23 | ** | 2.58 ± 0.14 | 2.52 ± 0.38 | |
| *Carboxylic acids* | 48 | 1.60 ± 0.34 | 2.01 ± 0.23 | * | 1.82 ± 0.18 | 1.79 ± 0.48 | |
| | 72 | 1.94 ± 0.14 | 2.25 ± 0.15 | ** | 2.07 ± 0.15 | 2.12 ± 0.28 | |
| | 96 | 2.10 ± 0.10 | 2.36 ± 0.12 | ** | 2.19 ± 0.14 | 2.27 ± 0.21 | |
| *Amines, amides, amino acids* | 48 | 0.92 ± 0.52 | 1.45 ± 0.12 | ** | 1.33 ± 0.30 | 1.04 ± 0.56 | |
| | 72 | 1.39 ± 0.17 | 1.65 ± 0.17 | ** | 1.51 ± 0.09 | 1.52 ± 0.30 | |
| | 96 | 1.52 ± 0.11 | 1.77 ± 0.18 | ** | 1.61 ± 0.08 | 1.67 ± 0.27 | |
| *Miscellaneous* | 48 | 0.96 ± 0.31 | 1.51 ± 0.26 | * | 1.21 ± 0.44 | 1.26 ± 0.39 | |
| (Unspecified) | 72 | 1.22 ± 0.38 | 1.90 ± 0.39 | * | 1.67 ± 0.53 | 1.45 ± 0.52 | |
| | 96 | 1.30 ± 0.44 | 1.95 ± 0.39 | * | 1.66 ± 0.53 | 1.60 ± 0.56 | |

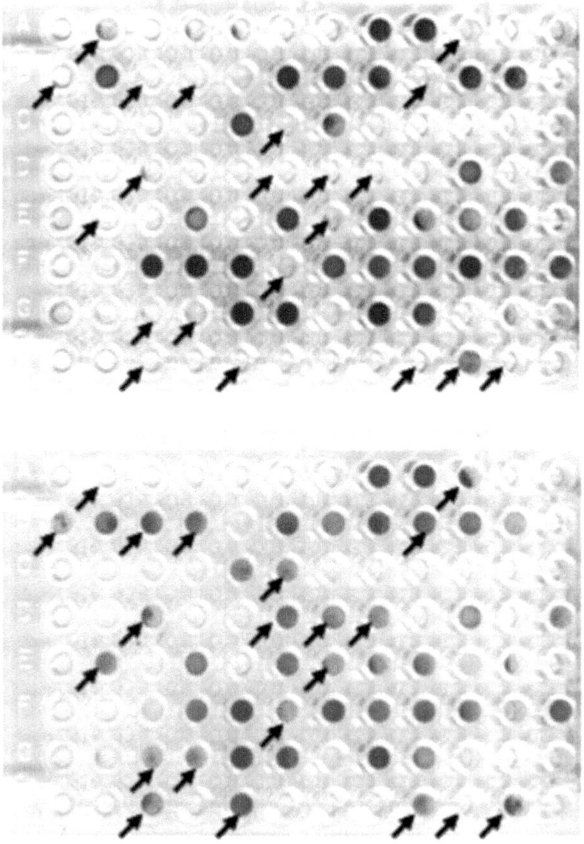

**Figure 1.** Example of Biolog GP2 Metabolic fingerprints of taproot (top) and of lateral root (bottom) rhizosphere bacterial communities. Arrows indicate qualitative differences in carbon source utilization between the two bacterial communities.

represented by the taproots. Out of these, four genera were represented only in the lateral roots RBCs, while only one genus was unique for the taproots.

Four selected bacterial strains were further identified by sequencing of their 16S rDNA genes. The identified isolates were *Herbaspirillum* sp., *Pseudomonas aeruginosa*, *Burkholderia cepacia* genomovar III and *Pseudomonas alclaigenes*.

*Herbaspirillum* spp was detected by Real-Time polymerase chain reaction (PCR) only on lateral roots of both control and salt treated plants but not on their taproots (Tab. 4). Relative to the total bacterial targets, the *Herbaspiriilum* spp populations comprised only 0.1% of the bacterial community under non-saline but 0.2% under saline conditions (Tab. 4). *P. aeruginosa* also distinguished between different root types and was found only on taproots (Tab. 4).

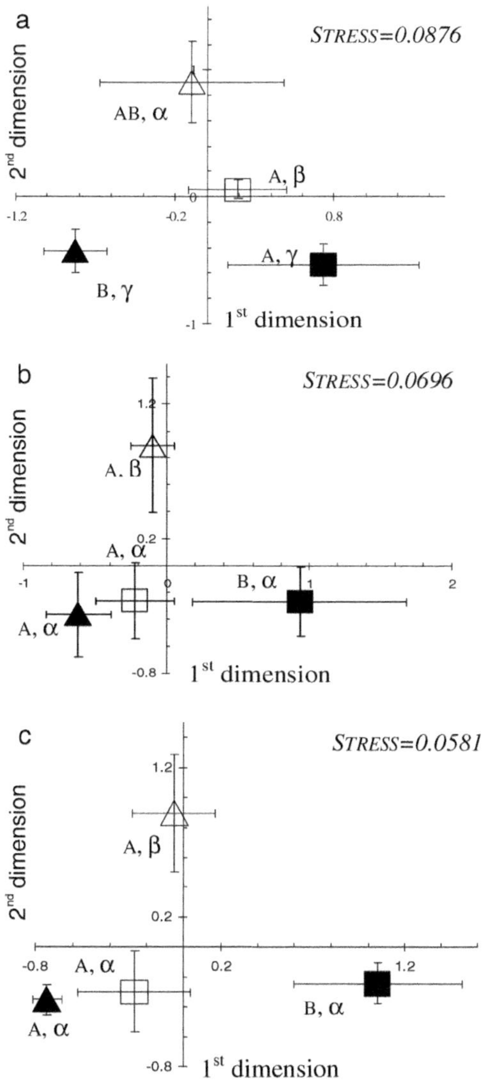


**Figure 2.** Non-metric multidimensional analysis (A, B and C) ordination of rhizosphere bacterial community (RBC) metabolic fingerprints in Biolog® GP2 microplates (in triplicates) after 48 (A), 72 (B) and 96 (C) hours of incubation. The different RBCs are marked: △ lateral roots of control plants; ▲ lateral roots of salt treated plants; □ taproots of control plants and ■ taproots of salt treatment plants. Stress values of the NMDA ordinations are indicated. Mean coordinates were compared by two-way ANOVA ($p < 0.05$). Different Latin letters indicate significant differences on the 1st dimension. Different Greek letters indicate significant differences on the 2nd dimension.

**Table 3.** Generic diversity of cultivable bacteria of *Vicia faba* lateral roots and taproots rhizosphere bacterial communities. Conspicuous pure culture isolates were identified using the Biolog® GN2 assay

| | Taproots | | | Lateral roots | | |
|---|---|---|---|---|---|---|
| | Total | Gram negative | Gram positive | Total | Gram negative | Gram positive |
| No. of isolates | 35(100%) | 30 (85.7%) | 5 (14.3%) | 71(100%) | 54 (76.1%) | 17 (23.9%) |
| Of which identified | 12(34.3%) | 12 | 0 | 21(29.6%) | 21 | 0 |
| No. of genus | 3 | 3 | 0 | 7 | 7 | 0 |
| Occurrence of genera | | | | | | |
| *Acidovorax* | | – | | | 1 | |
| *Agrobacterium* | | – | | | 2 | |
| *Burkholderia* | | 2 | | | 2 | |
| *Comamonas* | | – | | | 1 | |
| *Cytophaga* | | 4 | | | – | |
| *Herbaspirillum* | | – | | | 6 | |
| *Pseudomonas* | | 6 | | | 7 | |
| *Riemerella* | | – | | | 2 | |

**Table 4.** Real-Time polymerase chain reaction detection of Eubacteria (a), of *Herbaspirillum* spp. (b) and of *Pseudomonas aeruginosa* (c) in rhizosphere bacterial communities of taproots and lateral roots of faba beans (*Vicia faba* L.).

**a**

| Sample | Eubacteria<br>Slope: −5.2809; Intercept: 68.932; Coefficient: 0.99 | | | |
|---|---|---|---|---|
| | Log CFU $g^{-1}$ root f. wt | * Log targets $\mu l^{-1}$ sample | Log targets $g^{-1}$ root f. wt | Target $CFU^{-1}$ |
| LC | 9.66 | 6.59 | 11.94 | 190 |
| TC | 8.94 | 5.76 | 11.29 | 224 |
| LS | 8.56 | 6.07 | 11.50 | 871 |
| TS | 8.73 | 6.03 | 11.81 | 1202 |

**b**

| Sample | *Herbaspirillum*<br>Slope: −2.670; Intercept: 41.601; Coefficient: 0.94;<br>Detection limit: $10^1$ targets $\mu L^{-1}$ (cycle 39) | | | | |
|---|---|---|---|---|---|
| | Threshold cycle | *Targets $\mu l^{-1}$ sample | Log targets $g^{-1}$ root f. wt | % of Eubacteria | Melting peak |
| Positive control | 21 | $6.64 \times 10^7$ | | | 90.5 |
| LC | 34 | 505 | 7.93 | 0.010 | 89 |
| TC | 46 | 0 | 0.00 | 0.000 | 90.5 |
| LS | 35 | 327 | 7.82 | 0.021 | 89.5 |
| TS | 40 | 0 | 0.00 | 0.000 | 89.5 |

**c**

| Sample | *Pseudomonas aeruginosa*<br>Slope: −4.238; Intercept: 44.257; Coefficient: 0.987;<br>Detection limit: $10^1$ targets $\mu L^{-1}$ (cycle 39) | | | | |
|---|---|---|---|---|---|
| | Threshold cycle | *Targets $\mu l^{-1}$ sample | Log targets $g^{-1}$ root f. wt | % of Eubacteria | Melting peak |
| Positive control | 21 | $2.54 \times 10^5$ | | | 94.5 |
| LC | 41 | 5.9 | 0.00 | 0.000 | 93.5 |
| TC | 38 | 23 | 7.58 | 0.020 | 94.5 |
| LS | 40 | 10.8 | 0.00 | 0.000 | 93 |
| TS | 37 | 51 | 8.01 | 0.016 | 93.5 |

* Calculated according to the standard curve

## Discussion

The root system of faba bean plants (*Vicia faba* L.) is composed of inherently distinct roots that differ biochemically and physiologically [2]. We have hypothesized that such differences have some bearing on the assemblage of distinct bacterial communities associated with different roots.

Reciprocal effects between higher plants and microorganisms have been well demonstrated [18]. Rhizosphere microbial communities differ in abundance and in diversity from bulk soil communities [5, 19–21]. Diversity of rhizosphere microbial communities was shown to be plant-species dependent [5, 22–24]. Moreover, it has been demonstrated that the species composition of soil microbial communities may determine diversity of the higher plant associations, their productivity and their stability in a given ecosystem [25, 26].

Are such general differences expressed also at a finer level of resolution?

The results presented herein revealed differences between the metabolic fingerprints of rhizosphere bacterial communities of different roots and under different environments. Altogether the results prove that indeed, different roots are associated with distinct rhizosphere bacterial communities.

In order to verify the obtained results and to expand their scope, such distinction was further investigated under salt stress conditions. As with other legumes salinity significantly reduced the growth of *Vicia faba* plants, their nitrogen and protein content [27], and their roots' and leaves' ABA content [4, 28]. Salinity may also affect plant interactions with microorganisms [29]. Thus, to our results, the distinction between the different root types should consider not only the root type but also the environmental salinity.

Previous investigations have shown differences between bacterial communities associated with root segments of different developmental stages of *Cicer arietinum*, *Brassica napus* and *Sorghum bicolor* [5]. It was also found for roots of *Zea mays* [11], and even between cluster roots and non-cluster roots of soil grown *Lupinus albus* [13]. For soil grown plants of *Zea mays*, it was demonstrated that each segment was associated with different assortments of *Burkholderia cepacia* genotypes [30].

Real-Time PCR analysis showed definite root type specific colonization by *Herbaspirillum* spp. *Herbaspirillum* spp are plant growth promoting diazotrophs commonly associated with several gramineous plants [31–37], which were regarded as its 'natural hosts' [34]. Our results support the observations of Valverde et al. [38] and prove that *Herbaspirillum* can also be a common associate of legumes.

A major contribution of the present research is the fact that the results were obtained for roots that were grown under conditions that guarantied uniformity around each and every root of the root system [39]. By that we have proven that distinction between the bacterial communities of the rhizosphere results from inherent differences between the examined root types, and that such differences may be changed by environmental salinity.

## Conclusions

The root system of *Vicia faba* is comprised of inherently distinct root types that are associated by characteristic bacterial communities. Such bacterial communities differ not only by the root characteristics but also by the salinity of the environment.

## References

1 Marschner H, Römheld V (1996) Root-induced changes in the availability of nutrients in the rhizosphere. In: Waisel Y, Eshel A, Kafkafi U (eds.) *Plant Roots: The Hidden Half*, 2$^{nd}$ edition. New York, USA: Marcel Dekker 557–580

2 Lev R (2000) Comparison between the physiology of taproots and lateral roots of faba been (*Vicia faba*). PhD Thesis, Tel Aviv University, Israel

3 Hochholdinger F, Woll K, Sauer M, Dembinsky D (2004) Genetic dissection of root formation in maize (*Zea mays*) reveals root-type specific developmental programmes. *Annals of Botany* 93: 359–368
4 Waisel Y, Eshel A (2002b) Functional diversity of various constitutes of a single root system. In: Waisel Y, Eshel A, Kafkafi U (eds.) *Plant Roots: The Hidden Half*, 3rd edition. New York, USA: Marcel Dekker 157–173
5 Marschner P, Yang CH, Lieberie R, Crowley DE (2001) Soil and plant specific effect on bacterial community composition in the rhizosphere. *Soil Biology and Biochemistry* 33: 1437–1445
6 Kapulnik Y, Okon Y (2002) Plant growth promotion by rhizospheric bacteria. In: Waisel Y, Eshel A, Kafkafi U (eds.) *Plant Roots: The Hidden Half*, 3rd edition. New York, USA: Marcel Dekker 869–886
7 Rovira AD, Bowen GD, Foster RC (1983) The significance of mycoflora and mycorrhizas in plant nutrition. In: Lauchli A, Bielaski RL (eds.) *Encyclopedia of Plant Physiology, New Series, Vol 15A, Inorganic Plant Nutrition*. Springer Verlag, Berlin 61–93
8 Benizri E, Dedorouge O, Dibattista-Leboeuf C, Piutti S, Nguyen C, Guckert A (2002) Effect of rhizodeposits on soil microbial community structure. *Applied Soil Ecol* 21: 261–265
9 Hill GT, Mitkowske NA, Aldrich-Wolfe L, Jurkonie DD, Ficke A, Maldonado-Ramorez S, Lynch ST, Nelson EB (2000) Methods for assessing the composition and diversity of soil microbial communities. *Applied Soil Ecology* 15: 25–36
10 Baudoin E, Benizri E, Guckert A (2001) Metabolic fingerprint of microbial communities from distinct maize rhizosphere compartments. *Eur J Soil Biol* 37: 85–93
11 Baudoin E, Benizri E, Guckert A (2002) Impact of growth stage on the bacterial community structure along maize roots, as determined by metabolic and genetic fingerprinting. *Applied Soil Ecol* 19: 135–145
12 Kent AD, Triplett EW (2002) Microbial communities and their interactions in soil and rhizosphere ecosystems. *Ann Rev Microbiol* 56: 211–236
13 Marschner P, Neumann G, Kania A, Weiskopf L, Lieberei R (2002) Spatial and temporal dynamics of the microbial community structure in the rhizosphere of cluster roots of white lupin (*Lupinus albus* L.). *Plant and Soil* 246: 167–174
14 Hoagland DR, Arnon DI (1950) The water-culture method for growing plants without soil. *Califirnia Agriculture Experiment Station*, Circular no. 347. California, USA: University of California, Berkeley
15 Kuske CR, Ticknor LO, Miller ME, Dunbar JM, Davis JA, Barns SM, Belnap J (2002) Comparison of soil bacterial communities in rhizospheres of three plant species and the interspaces in an arid grassland. *Appl Environ Microbiol* 68: 1854–1863
16 Garland JL, Mills AL (1991) Classification and characterization of heterotrophic microbial communities on the basic patterns of community-level sole-carbon-source utilization. *Appl Environ Microbiol* 57: 2351–2359
17 Derry AM, Staddon WJ, Kevan PJ, Trevors JT (1999) Functional diversity of microorganisms in three arctic soils as determined by sole-carbon-source-utilization. *Biodiversity and Conservation* 8: 205–221
18 Ruppel S (2000) Role of rhizosphere and endophytic bacteria in plant nutrition. *Arch Agronomy and Soil Sci* 45: 329–341
19 Marilley L, Vogt G, Blanc M, Aragno M (1998) Bacterial diversity in the bulk soil and rhizosphere fractions of *Lolium perenne* and *Trifolium repens* as revealed by PCR restriction analysis of 16S rDNA. *Plant and Soil* 198: 219–224

20 Smalla K, Weiland G, Buchner A, Zock A, Parzy J, Kaiser S, Roskot N, Heuer H, Berg G (2001) Bulk and rhizosphere soil bacterial communities studied by denaturing gradient gel electrophoresis: Plant-dependent enrichment and seasonal shifts revealed. *Appl Environ Microbiol* 67: 4742–4751
21 Kandeler E, Marchner P, Tscherko D, Gahoonia TS, Nielsen NE (2002) Microbial community composition and functional diversity in the rhizosphere of maize. *Plant and Soil* 238: 301–312
22 Kowalchuk GA, Buma DS, de Boer W, Klinkhamer PGL, van Veen JA (2002) Effects of above-ground plant species composition and diversity on the diversity of soil-borne microorganisms. *Antoni van Leewenhoek International Journal of General and Molecular Microbiology* 81: 509–520
23 Johnson MJ, Lee KY, Scow KM (2003) DNA fingerprinting reveals links among agricultural crops, soil properties, and the composition of soil microbial communities. *Geoderma* 114: 279–303
24 Miethling R, Ahrends K, Tebbe CC (2003) Structural differences in the rhizosphere communities of legumes are not equally reflected in community-level physiological profiles. *Soil Biol and Biochem* 35: 1405–1410
25 Van der Heijden MGA, Klironomos JN, Ursic M, Moutoglis P, Streitwolf-Engel R, Boller T, Wiemken A, Sanders IR (1998) Mycorrhizal fungal diversity determines plant biodiversity, ecosystem variability and productivity. *Nature* 396: 69–72
26 Van der Heijden MGA, Wiemken A, Sanders IR (2003) Different arbascular mycorrhizal fungi alter coexistence and resource distribution between co-occurring plants. *New Phytologist* 157: 569–578
27 Delgado MJ, Ligero F, Lluch C (1993) Effects of salt stress on growth and nitrogen fixation by pea, faba-bean, common bean and soybean plants. *Soil Biol and Biochem* 26: 371–376
28 Degenhardt B, Gimmler H, Hose E, Hartung W (2000) Effect of alkaline and saline substrates on ABA content, distribution and transport in plant roots. *Plant and Soil* 225: 83–94
29 Tal O (1993) The expression of *Rhizobium meliloti* nodABC genes in response to alfalfa flavonoids and root exudates under saline conditions. MSc Thesis. The Hebrew University of Jerusalem, Israel
30 Chiarini L, Giovannelli V, Bevivino A, Dalmastri C, Tabacchioni S (2000) Different portions of the maize root system host *Burkholderia cepacia* populations with different degrees of genetic polymorphism. *Environ Microbiol* 2: 111–118
31 Baldani VLD, Baldani JI, Doberiner J (2000) Inoculation of rice plants with the endophytic diazotrophs *Herbaspirillum seropedica* and *Burkholderia* spp. *Biol and Fertility of Soils* 30: 485–491
32 Elbeltagy A, Nishioka K, Sato T, Suzuki HYB, Hamada T, Isawa T, Mitsui H, Minamisawa K (2001) Endophytic colonization and in planta nitrogen fixation by a *Herbaspirillum* sp. isolated from wild rice species. *Appl Environ Microb* 67: 5285–5293
33 Gynaeshwar P, James EK, Reddy PM, Ladha J (2002) *Herbaspirillum* colonization increases growth and nitrogen accumulation in aluminium-tolerant rice varieties. *New Phytologist* 154: 131–145
34 James EK, Gyaneshwar P, Mathan N, Barraquio WL, Reddy PM, Iannetta PPM, Olivares FL, Ladha JK (2002) Infection and colonization of rice seedlings by the plant growth-promoting bacterium *Herbaspirillum seropedica* Z67. *Molecular Plant-Microbe Interactions* 15: 894–906
35 Oliviera ALM, Urquiaga S, Dobereiner J, Baldani JI (2002) The effect of inoculating endophytic $N_2$-fixing bacteria on micropropagated sugarcane plants. *Plant and Soil* 242: 205–215

36 Boddey RM, Urquiaga S, Alves BRJ, Reis V (2003) Endophytic nitrogen fixation in sugarcane: present knowledge and future applications. *Plant and Soil* 252: 139–149

37 Roncato-Maccari LDB, Ramos HJO, Pedrosa FO, Alquini Y, Chubatsu LS, Yates MG, Rigo LU, Steffens MBR, Souza EM (2003) Endopytic *Herbaspirillum seropedica* expresses *nif* genes in gramineous plants. *FEMS Microbiol Ecol* 45: 39–47

38 Valverde A, Velázquez E, Gutiérrez C, Cervantes E, Ventosa, Igual JM (2003) *Herbaspirillum lustianum* sp. nov., a novel nitrogen-fixing bacterium associated with root nodules of *Phaseolus vulgaris*. *Int J Systematic Microbiol* 53: 1976–1983

39 Waisel Y, Eshel A (2002a) Aeroponics: A tool for root research under minimal environmental restriction. In: Waisel Y, Eshel A, Kafkafi U (eds.) *Plant Roots: The Hidden Half*, 3$^{rd}$ edition. New York, USA: Marcel Dekker 323–331

Biosaline Agriculture and Salinity Tolerance in Plants
Edited by M. Öztürk, Y. Waisel, M.A. Khan and G. Görk


# Tolerance of some potential forage grasses from arid regions of Pakistan to salinity and drought

Muhammad Ashraf

*Department of Botany, University of Agriculture, Faisalabad 38040, Pakistan*

## Introduction

Pakistan falls into arid and semi-arid regions, as about an area of 562,592 $km^2$ out of 803,935 $km^2$ of the total area of Pakistan is the arid land with an annual rainfall of less than 60 cm. Of the total salinized land, 56,656 $km^2$ are saline, 29,138 $km^2$ are saline sodic and 283 $km^2$ sodic soil [1]. A large part of available agricultural land remains non-irrigated due to shortage of water. The unavailability of water and low rainfall are the major factors for converting large areas into deserts.

The four main deserts of Pakistan are Thal and Cholistan in the Punjab, Thar in Sindh and Kharan in Baluchistan. Cholistan and Thar are the largest, with annual rainfall of less than 12 cm. The major problems in all deserts are the scarcity of water for irrigation and escalating levels of salts in the soil. The subsoil water in most of the places is brackish.

The twin menace, i.e., salinity and drought, has caused many social and economic problems in the area such as poor living standards, health problems for animals and humans. Selection of salt and drought resistant plant species and their domestication on salt and drought-hit areas is very economic and feasible for overcoming drought and salinity problems [2–4].

Grasses such as *Cenchrus pennisetiformis, Leptochloa fusca, Panicum turgidum* and *Pennisetum divisum* have great importance as forage for livestock and dairy development in deserts. In view of the great economic importance and nutritional value of these grasses, the present study was undertaken to determine the relative resistance to salinity and drought. The major objective of the study was to identify highly salt and drought tolerant species, which may thrive in arid regions under irrigation with subsoil brackish water.

## Materials and methods

In the drought experiment, seeds of *Cenchrus pennisetiformis, Panicum turgidum* and *Pennisetum divisum* were collected from the Cholistan and Thal deserts, whereas those of *Leptochloa fusca* were collected from a derelict field. Drought cycles were started after about 6-week normal growth. The drought treatments maintained during the experiment were control (watering daily to field capacity), mild stress (plants were droughted four times until wilting and re-watered to field capacity), and severe stress (plants were droughted eight times as in mild stress). Various growth, biochemical, and physiological parameters, were recorded during the experiment to evaluate different grasses against drought.

In the salinity experiment, the same grasses as in the drought experiment were used. *Puccinellia distans* was included in this study as it is known for its high salt tolerance. The seed of *Puccinellia distans* was obtained from England. All seed samples were sown in Petri dishes. After 2 weeks the seedlings were transplanted into 18 cm plastic pots containing sand. The salt treatments used were 0, 8, 16 and 24 dS/m, that were prepared by mixing four salts, $NaHCO_3$, $MgSO_4.7H_2O$, $CaCl_2.2H_2O$ and NaCl in a ratio of 1:5:10:30 in half strength Hoagland nutrient solution.

**Leaf hydration (H):** It is the ratio of the weight of water in a turgid leaf to its dry weight

$H = W_{ref}/W_d$

Where, $W_{ref}$ = Weight in leaf at the reference, $W_d$ = dry weight of the leaf

**Leaf elasticity** was calculated as:

$E\ (MPa) = (\psi_{w2} - \psi_{w1})/\Delta R$

$\Delta R = (W_1 - W_2)/W_{ref}$

$W_{ref} = W_1 - W_d + (\psi_{w1} - 0.5)\ (W_1 - W_2)/(\psi_{w2} - \psi_{w1})$

Where, $W_{ref}$ = Weight of water in a leaf at the reference, $\Delta R$ = change in relative water content, $W_1$ = Initial weight of leaf, $W_2$ = Final weight of leaf, $W_d$ = Dry weight of leaf, $\Psi_w$ = water potential of the leaf when over-pressed to –0.5 MPa, $\Psi_{w1}$ = Water potential of the same leaf measured after over-pressing and weighing, $\Psi_{w2}$ = Water potential of the same leaf after over-pressing for 90 seconds.

**Osmotic adjustment** was calculated by finding the difference between osmotic potential of rehydrated and control plants.

Methods used for the determination of different physiological and biochemical parameters are presented in Table 1.

**Table 1.** Methods used for the determination of various physiological parameters

| **Parameter** | **Method used** |
|---|---|
| Proline estimation | Bates et al. [5] |
| Chlorophyll estimation | Witham et al. [6] |
| Wax content | Silva Fernandes et al. [7] |
| Total soluble proteins | Lowry et al., 1951 [8] |
| Leaf resistance | By porometer (MK3 Delta-T) |
| Osmotic potential | By vapor pressure osmometer |
| Analysis of ions | By flame photometer |
| Leaf elasticity | Thomas [9] |

## Result and discussion

*Drought experiment*

Increasing drought cycles had an adverse effect on fresh and dry matter production in all four species (Tab. 2, Fig. 1). Shoot fresh and dry weights in all species decreased consistently at both drought treatments. *Cenchrus pennisetiformis* and *Panicum turgidum* had significantly greater fresh and dry biomass compared with the other species.

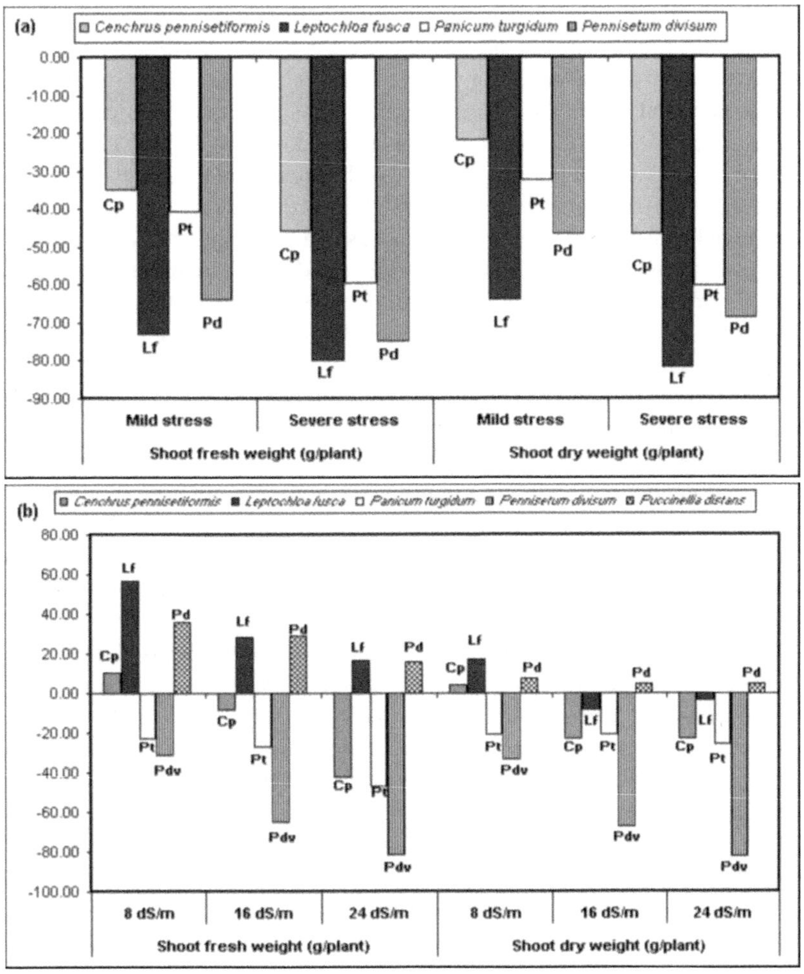


**Figure 1.** Percent increase or decrease with respect to control of shoot fresh and dry weight of some grasses from Cholistan desert under (a) drought and (b) salinity stress

Thus, *C. pennisetiformis* and *P. turgidum* were relatively resistant to the repeated cycles of drought. In contrast, the lower fresh weight and dry mass production of *L. fusca* shows its susceptibility to drought stress. The relatively greater drought resistance of *C. pennisetiformis* and *P. turgidum* is expected in view of the fact that both species are natural colonizers of the desert area where severe water deficit conditions are predominant [10]. The susceptibility of *L. fusca* to drought is also expected as the species commonly occurs in waterlogged sodic soils [1].

Leaf solute potential decreased significantly in all four grasses with increase in drought cycles. *Leptochloa fusca* had the lowest and *P. divisum* the highest osmotic potential among all the grasses at both drought treatments (Tab. 2). Generally, leaf osmotic potential was more negative in plants which had experienced wilting. This clearly showed that osmotic adjustment was generally increased in plants of all species, which experienced drought cycles. *Leptochloa fusca* had significantly lower osmotic potential but higher osmotic adjustment than the other grasses. Osmotic adjustment plays a central role in plant adaptation to drought [11, 12], although it is known to be effective in plant tolerance to other stress environments such as salinity and freezing [13]. It is now widely accepted that a decline in osmotic potential at full turgor may be due to the apoplastic as well as to accumulation of osmotica [14]. The lower osmotic potentials of plants experienced drought treatments compared with those of plants, which were well watered, may have been due to the accumulation of osmotically active solutes. It thus indicates that osmotic adjustment occurred in all species subjected to drought.

The extent of osmotic adjustment was significantly higher in *L. fusca* and *P. divisum* than the other species (Tab. 2). The results for osmotic adjustment are surprising in view of their contrasting capacity to resist moisture stress. However, there might be some other attributes, which have been contributed to enhance drought resistance in *C. pennisetiformis* and *P. turgidum*. Leaf diffusive resistance in *C. pennisetiformis* which may have been a major factor for controlling water loss, as both *C. pennisetiformis* and *P. turgidum* maintained a non-wilted condition to considerably lower water potential.

Leaf resistance of *C. pennisetiformis* and *P. divisum* increased with increase in drought cycles at morning and evening, whereas that of *L. fusca* and *P. turgidum* remained constant. *Pennisetum divisum* had significantly greater leaf resistance at noon compared with the other three species at 4 drought cycles. Leaf resistance of all the grasses except *C. pennisetiformis* decreased consistently with the increase in drought cycles in evening (Tab. 2).

Cell wall elasticity of *C. pennisetiformis, P. turgidum* and *P. divisum* decreased with the increase in drought cycles, whereas that of *L. fusca* remained unaffected. *Pennisetum divisum* had the lowest and *L. fusca* the highest elasticity among all the species at 4 and 8 drought cycles, respectively (Tab. 2). The elasticity of the cell wall allows volume changes to occur over a range of hydrostatic pressures [15]. The elasticity is an important parameter in cell wall relations, controlling the manner in which cell water potential changes as the cell volume changes [16]. In the present studies, however, drought cycles had no significant effect on elasticity of drought sensitive *Leptochloa fusca,* while it was decreased in other species.

**Table 2.** Growth, physiological and biochemical parameters of some grasses subjected to mild or sever drought stress

| Grass species | Shoot fresh weight (g/plant) | | | Shoot dry weight (g/plant) | | | Relative water content (%) | | |
|---|---|---|---|---|---|---|---|---|---|
| | Control | Mild stress | Severe stress | Control | Mild stress | Severe stress | Control | Mild stress | Severe stress |
| *Cenchrus pennisetiformis* | 20.30 d | 13.20 a | 11.00 a | 9.70 b | 7.60 ab | 5.20 a | 71.40 b | 70.30 a | 87.30 a |
| *Leptochloa fusca* | 25.30 b | 6.80 b | 5.10 b | 11.40 a | 4.10 c | 2.10 b | 45.80 c | 71.00 a | 56.60 b |
| *Panicum turgidum* | 23.70 c | 14.10 a | 9.60 a | 12.10 a | 8.20 a | 4.80 a | 69.40 b | 70.50 a | 52.40 b |
| *Pennisetum divisum* | 31.75 a | 11.50 a | 8.00 a | 11.50 a | 6.15 b | 3.63 ab | 90.65 a | 86.30 a | 81.70 a |
| | Epicuticular wax content ($\mu g/cm^2$) | | | Leaf osmotic potential (-MPa) | | | Osmotic adjustment | | |
| *Cenchrus pennisetiformis* | 61.70 bc | 158.50 c | 153.20 c | 1.40 a | 2.30 b | 2.50 b | | 0.91 | 1.11 |
| *Leptochloa fusca* | 34.50 c | 281.80 a | 321.90 a | 1.10 a | 2.90 a | 3.20 a | | 1.80 | 2.12 |
| *Panicum turgidum* | 97.70 b | 209.20 b | 118.70 d | 1.35 a | 2.03 b | 2.30 bc | | 0.70 | 0.95 |
| *Pennisetum divisum* | 185.02 a | 183.65 c | 222.40 b | 0.44 a | 1.65 c | 2.00 c | | 1.21 | 1.50 |
| | Leaf resistance(s/cm) | | | | | | | | |
| | Morning | | | Noon | | | Evening | | |
| *Cenchrus pennisetiformis* | 4.10 a | 7.30 ab | 15.40 a | 3.10 b | 5.00 b | 5.60 a | 12.60 a | 10.50 a | 13.00 a |
| *Leptochloa fusca* | 4.80 a | 6.30 b | 7.00 c | 3.81 b | 3.10 b | 3.90 a | 10.80 a | 9.60 ab | 6.00 b |
| *Panicum turgidum* | 5.50 a | 4.64 b | 5.70 c | 3.55 b | 3.70 b | 3.54 a | 10.90 a | 8.00 ab | 7.10 b |
| *Pennisetum divisum* | 3.05 a | 11.50 a | 11.02 b | 11.65 a | 13.00 a | 5.00 a | 13.65 a | 7.00 b | 8.00 b |
| | Leaf elasticity (MPa) | | | Leaf hydration (g water/g) | | | Proline (μmol/g fresh weight) | | |
| *Cenchrus pennisetiformis* | 6.15 ab | 6.20 b | 9.90 a | 3.60 b | 3.00 b | 2.90 b | 1.65 ab | 3.90 b | 5.60 b |
| *Leptochloa fusca* | 5.30 b | 5.30 b | 6.70 b | 5.00 a | 4.40 a | 2.70 b | 3.65 a | 6.30 a | 8.30 a |
| *Panicum turgidum* | 5.80 b | 7.60 b | 10.32 a | 3.90 b | 3.70 ab | 3.40 a | 2.04 ab | 3.90 b | 3.40 c |
| *Pennisetum divisum* | 8.60 a | 12.14 a | 12.90 a | 3.90 b | 3.40 ab | 2.40 b | 1.01 b | 1.00 c | 2.95 c |
| | Chlorophyll a (mg/g fresh weight) | | | Chlorophyll b (mg/g fresh weight) | | | Total chlorophyll (mg/g fresh weight) | | |
| *Cenchrus pennisetiformis* | 1.25 | 1.35 | 1.35 | 0.82 | 0.74 | 0.74 | 2.07 | 2.09 | 2.09 |
| *Leptochloa fusca* | 1.30 | 1.30 | 1.35 | 1.00 | 1.00 | 1.00 | 2.30 | 2.30 | 2.35 |
| *Panicum turgidum* | 1.23 | 1.20 | 1.20 | 0.70 | 0.80 | 0.80 | 1.93 | 2.00 | 2.00 |
| *Pennisetum divisum* | 1.10 | 1.10 | 1.10 | 0.60 | 0.70 | 0.70 | 1.70 | 1.80 | 1.80 |
| | Total soluble proteins (mg/g fresh weight) | | | | | | | | |
| *Cenchrus pennisetiformis* | 1.20 a | 1.10a | 1.03 b | | | | | | |
| *Leptochloa fusca* | 1.20 a | 1.00 a | 1.60 a | | | | | | |
| *Panicum turgidum* | 1.10 a | 1.00 a | 1.00 b | | | | | | |
| *Pennisetum divisum* | 1.00 a | 0.60 b | 1.00 b | | | | | | |

Increasing drought cycles significantly decreased the leaf hydration of all the four grasses. *Leptochloa fusca* and *P. turgidum* had the highest leaf hydration at 4 and 8 drought cycles, respectively. The data for relative water content of *C. pennisetiformis* and *L. fusca* increased, whereas it decreased at the highest drought treatment. In contrast, relative water content of *P. divisum* remained unaffected at both drought treatments (Tab. 2).

Increasing drought cycles had no significant effect on chlorophyll a, b, and total chlorophyll content. Species differed significantly for chlorophyll a and b, but this difference was attributed to large differences in their controls. Total soluble proteins of *L. fusca* increased at 8 drought cycles, whereas that of *P. divisum* decreased at 4 drought cycles. In contrast, protein content of *C. pennisetiformis* and *P. turgidum* remained unaffected at both drought treatments (Tab. 2). No significant effect on chlorophyll content in all four grasses may have been to the fact that the specific enzyme responsible for the synthesis of green pigment [17] was not affected by drought. The stability of chlorophyll may also be attributed to the greater accumulation of proline (Tab. 2) and other osmotica. Proline is known to help plants in tolerance of abiotic stresses [18, 19]. Proline forms a hydration sphere around macromolecules and protects them from denaturing under stressful conditions [20–22].

Proline content for *C. pennisetiformis* and *L. fusca* increased significantly, whereas that of *P. turgidum* and *P. divisum* remained unaffected at both drought treatments. *Leptochloa fusca* synthesized significantly greater amount of proline at both treatments compared with the other species (Tab. 2). Leaf proline content generally increased with increase in drought intensity, but greater increase was recorded in drought susceptible *L. fusca*. The results clearly show that proline levels in the drought sensitive species were inversely correlated with the ability to withstand severe drought.

Leaf epicuticular wax of *L. fusca* and *P. turgidum* increased consistently with increase in drought intensity, whereas that of *C. pennisetiformis* increased only at the mild stress level. Leaf wax content of *P. divisum* remained unaffected at both drought treatments (Tab. 2). Deposition of wax on leaf surface plays a crucial role in minimizing water loss [23]. Although all the plants accumulated considerable quantity of wax under drought conditions, *L. fusca* and *P. divisum* showed the highest epicuticular wax content, which indicates a negative correlation to drought resistance. However, Ashraf and Mehmood [24] reported a positive correlation between deposition of wax on leaf surface and drought resistance in some *Brassica* species.

On the basis of these results it can be concluded that *C. pennisetiformis* and *P. turgidum* were the most drought tolerant, *P. divisum* being intermediate and *L. fusca* the most sensitive to drought stress. However, very few relationships were observed between the drought tolerance of these grass species and different physiological attributes examined in this study.

### *Salinity experiment*

The five grass species *Leptochloa fusca, Cenchrus pennisetiformis, Panicum turgidum, Pennisetum divisum* and *Puccinellia distans* responded differently to increasing salinity level of rooting medium. Increasing salinity treatment markedly reduced shoot and root dry weights in *C. pennisetiformis, P. turgidum* and *P. divisum*, whereas those of *L. fusca* and *P. distans* showed stability at all salinity treatments (Tab. 3). *Leptochloa fusca* had significantly greater shoot fresh and dry matter than the other grass species at 16 and 24 dS/m. Percent fresh and dry matters of *L. fusca* and *P. distans* remained unaffected at all salt treatments (Tab. 3). In contrast, percent biomass production in *P. divisum* was severely reduced by salinity (Fig. 1). The results for the biomass production clearly shows that *L. fusca* and *P. distans* were highly tolerant to varying salinity levels of the growth medium compared with the other three species. The better performance of these two species is expected as they were already found to be highly salt tolerant, e.g., *L. fusca*[1], and *P. distans* [2]. *Pennisetum divisum* was the poorest of all species, whereas *C. pennisetiformis* and *P. turgidum* were intermediate in performance in response to salinity.

Leaf osmotic potential of all five species except *L. fusca* decreased with increase in salt concentration. *Puccinellia distans* and *P. turgidum* had the lowest osmotic potential at 8 and 16 dS/m, respectively, whereas *L. fusca* had the highest osmotic potential at 24 dS/m of all the five species. The remaining species did not differed significantly at all salt treatments (Tab. 3).

Salinity treatments had no affect on leaf soluble proteins of all the grass species while proline content increased with increasing salinity treatments. Proline content of *L. fusca* and *P. distans* increased consistently with increasing stress level whereas, that of *C. Pennisetiformis*, *P. turgidum*, and *P. divisum* remained unaffected at all the treatments. *Puccinellia distans* had significantly greater amount of proline compared with all the other species at all salt treatments. *Leptochloa fusca* also synthesized greater quantity of proline than the remaining three species at the higher salt treatments. The proline content of the remaining species was almost similar at all salt treatments (Tab. 3).

The high osmotic potential of *L. fusca* are less easy to explain in view of its considerable salt tolerance and high accumulation of proline in the leaves, as proline has been reported to be osmotically very active [25]. However, there was a positive correlation between low osmotic potential and high proline content in *P. distans*. In contrast, the low osmotic potential values in the other three species cannot be explained in view of the organic solutes analyzed in this study. It is possible that sugars, organic acids or other quaternary ammonium compounds that have not been determined in this study might have played a role in maintaining low osmotic potential.

Increasing salt concentration in the rooting medium had no effect on relative water content of all the five grass species. The maximum water content of all grasses was observed at 16 dS/m while it was considerably lower in other treatments. *Cenchrus pennisetiformis* showed the maximum water content at this treatment followed by *P. divisum* and *P. turgidum* (Tab. 3).

The mean data for chlorophyll (Tab. 3) showed that increasing salt intensity had no effect on chlorophyll a, b, and total chlorophyll. *Leptochloa fusca* and *P. divisum* had higher Chl. b than the other species at 8 dS/m. In contrast, at 24 dS/m *L. fusca* and *P. turgidum* had greater Chl. b compared with the other species. *Leptochloa fusca*

**Table 3.** Effect of salinity on some growth, physiological and biochemical parameters of some grasses

| Grass species | Shoot fresh weight (g/plant) | | | | Shoot dry weight (g/plant) | | | | Root dry weight (g/plant) | | | |
|---|---|---|---|---|---|---|---|---|---|---|---|---|
| | | dS/m | | | | dS/m | | | | dS/m | | |
| | Control | 8 | 16 | 24 | Control | 8 | 16 | 24 | Control | 8 | 16 | 24 |
| *Cenchrus pennisetiformis* | 21.6 bc | 23.9 ab | 19.8 ab | 12.5 b | 9.2 b | 9.6 a | 7.1 ab | 7.1 ab | 5.3 ab | 4.7 bc | 5.2 b | 3.0 b |
| *Leptochloa fusca* | 18.2 c | 28.5 a | 23.4 a | 21.2 a | 8.8 b | 10.3 a | 8.1 a | 8.5 a | 3.6 bc | 3.1cd | 2.2 c | 1.7 c |
| *Panicum turgidum* | 27.7 b | 21.4 bc | 20.3 ab | 14.8 b | 9.3 b | 7.4 a | 7.4 ab | 6.9 ab | 7.2 a | 8.1 a | 10.5 a | 7.1 a |
| *Pennisetum divisum* | 44.3 a | 30.5 a | 15.7 b | 8.2 b | 16.5 a | 11.0 a | 5.5 b | 3.0 c | 6.0 a | 5.5 b | 2.2 c | 2.3 b |
| *Puccinellia distans* | 11.5 d | 15.6 c | 14.8 b | 13.3 b | 4.0 c | 4.3 b | 4.2 c | 4.2 b | 1.9 c | 2.7 d | 1.3 c | 1.8 bc |
| | Leaf resistance (s/cm) | | | | | | | | | | | |
| | Morning | | | | Noon | | | | Evening | | | |
| *Cenchrus pennisetiformis* | 5.2 a | 11.3 a | 7.0 a | 5.9 b | 2.2 a | 2.9 a | 6.2 b | 4.0 b | 9.7 a | 9.4 a | 10.1 b | 15.4 b |
| *Leptochloa fusca* | 5.6 a | 3.9 a | 4.8 a | 4.8 b | 4.6 a | 3.8 a | 3.2 b | 4.2 b | 5.7 a | 6.7 a | 11.8 b | 14.6 b |
| *Panicum turgidum* | 6.1 a | 6.9 a | 5.6 a | 1.2 b | 2.9 a | 2.5 a | 4.3 b | 4.1 b | 7.3 a | 6.3 a | 9.6 b | 11.5 b |
| *Pennisetum divisum* | 3.9 a | 7.4 a | 10.4 a | 15.5 a | 4.2 a | 2.3 a | 16.2 a | 13.1 a | 9.8 a | 12.3 a | 30.1 a | 36.0a |
| *Puccinellia distans* | 3.3 a | 2.6 a | 3.8 a | 2.7 b | 2.0 a | 1.8 a | 2.8 b | 1.7 b | 3.7 a | 4.3 a | 8.9 b | 3.4 b |
| | Leaf osmotic potential(-MPa) | | | | Relative water content (%) | | | | Epicuticular wax content ($\mu g/cm^2$) | | | |
| *Cenchrus pennisetiformis* | 0.8 a | 2.4 b | 2.4 b | 5.0 a | 72.1 | 61.4 | 80.4 | 67.2 | 37.0 d | 113.9 c | 102.3 b | 219.1 b |
| *Leptochloa fusca* | 0.9 a | 1.8 b | 2.0 b | 2.0 c | 52.1 | 53.9 | 77.6 | 49.3 | 248.2 b | 252.7 b | 207.0 a | 301.0 a |
| *Panicum turgidum* | 0.9 a | 1.7 b | 5.5 a | 5.3 a | 65.4 | 61.3 | 75.1 | 62.1 | 73.5 d | 126.7 c | 263.1 a | 87.1 c |
| *Pennisetum divisum* | 0.4 a | 0.9 b | 2.0 b | 4.0 b | 66.2 | 66.3 | 78.5 | 50.7 | 139.3 c | 318.5 a | 84.4 b | 80.0 c |
| *Puccinellia distans* | 1.6 a | 3.3 a | 2.3 b | 3.6 b | 64.8 | 57.2 | 78.3 | 58.5 | 321.6 a | 167.8 bc | 233.6 a | 289.1 a |
| | Chlorophyll a (mg/g f. wt.) | | | | Chlorophyll b (mg/g f. wt.) | | | | Total chlorophyll (mg/g f. wt.) | | | |
| *Cenchrus pennisetiformis* | 0.90 | 1.3 | 1.30 | 1.20 | 0.45 b | 0.70 c | 0.90 a | 0.50 c | 1.40 c | 2.30 c | 2.30 b | 1.80 c |
| *Leptochloa fusca* | 1.40 | 1.40 | 1.35 | 1.30 | 1.20 a | 1.30 a | 1.13 a | 1.40 a | 3.00 a | 3.30 a | 3.00 a | 2.70 a |
| *Panicum turgidum* | 1.30 | 1.30 | 1.33 | 1.30 | 0.90 a | 1.00 b | 0.90 a | 1.11 b | 2.50 b | 2.60 bc | 2.60 ab | 2.80 a |
| *Pennisetum divisum* | 1.10 | 1.05 | 1.20 | 1.20 | 0.60 b | 0.50 c | 0.60 b | 0.60 c | 1.75 c | 1.70 d | 2.10 b | 2.10 bc |
| *Puccinellia distans* | 1.30 | 1.30 | 1.30 | 1.30 | 1.30 a | 1.10 ab | 1.10 a | 1.10 b | 3.00 a | 2.90 ab | 2.60 a | 2.34 ab |
| | Total soluble proteins (mg/g f. wt.) | | | | Proline ($\mu$mol/g f. wt) | | | | | | | |
| *Cenchrus pennisetiformis* | 0.9 | 1.2 | 0.9 | 1.0 | 3.3 b | 3.2 b | 1.6 b | 2.9 c | | | | |
| *Leptochloa fusca* | 1.0 | 1.1 | 1.0 | 1.1 | 2.6 b | 2.6 b | 10.0 b | 33.4 b | | | | |
| *Panicum turgidum* | 0.9 | 1.2 | 1.1 | 1.3 | 3.0b | 3.0 b | 2.7 b | 3.2 c | | | | |
| *Pennisetum divisum* | 1.1 | 1.3 | 0.9 | 0.9 | 1.4 b | 2.4 b | 1.6 b | 4.7 c | | | | |
| *Puccinellia distans* | 1.4 | 1.1 | 1.2 | 1.3 | 38.5 a | 47.2 a | 83.6 a | 83.6 a | | | | |

also had greater total chlorophyll at all salt treatments, whereas *P. divisum* and *C. pennisetiformis* had lower total chlorophyll than the other species at 8 and 24 dS/m, respectively. No significant effect on chlorophyll content in all five grasses may have been due to specific enzyme responsible for the synthesis of green pigment [17]. The stability of chlorophyll may also be attributed to the greater accumulation of proline (Tab. 3) and other osmotica. Proline is known to help plants in tolerance of abiotic stresses [18, 19]. Proline forms a hydration sphere around macromolecules and protects them from denaturing under stressful conditions [20–22].

The leaf diffusive resistance of *P. divisum* determined in the morning increased consistently with the increase in the salt treatment, whereas those of all other four species remained almost consistent. *Pennisetum divisum* had highest leaf resistance of all five species in the morning at 24 dS/m. Leaf resistance of *P. divisum* measured at noon increased with increasing salt treatments, whereas those of other four species remained almost unchanged at all salt treatments except *C. pennisetiformis* that had significantly greater leaf resistance at 16 dS/m than its control. *Pennisetum divisum* had the highest, whereas *P. distans* the lowest leaf resistance at 16 and 24 dS/m. The leaf resistance of *L. fusca* and *P. divisum* increased, whereas those of *C. Penniseti-formis, P. turgidum and P. distans* remained unaffected at all salt treatments in the evening (Tab. 3).

A consistent increase in deposition of wax on leaf surface was observed in *L. fusca* and *P. distans* with increase in salt concentration of the growth medium. *Pennisetum divisum* was the lowest and the other two species intermediate in the deposition of wax. The results for epicuticular wax content of the five species can easily be correlated with their tolerance to salt stress. It is now well evident that epicuticular wax content on leaf epidermis plays a pivotal role in minimizing evaporative loss and thus maintaining high turgor [23]. However, the results for epicuticular wax cannot be correlated with the leaf resistance data of the five species, because *P. divisum* with low wax deposition had considerably high leaf resistance at the higher salinity treatments. In contrast, *P. distans* with excessive epicuticular wax was the lowest in leaf resistance among all species (Tab. 3).

Increasing external salt concentration affected shoot $Na^+$ concentration non-significantly (Tab. 4). *Cenchrus pennisetiformis* and *P. divisum* had significantly higher and *L. fusca* and *P. distans* lower shoot $Na^+$ at 16 dS/m, but at higher salinity (24 dS/m the difference in $Na^+$ concentrations in all the species was non-significant. Root $Na^+$ of all five species increased with increase in salt level. *Panicum turgidum* had significantly higher concentration than the other species at all salt treatments. At 8 dS/m *P. divisum* and *L. fusca* had intermediate and *C. pennisetiformis* and *P. distans* the lowest root $Na^+$.

$Cl^-$ concentrations in the shoots and roots in all five species increased with increase in salt treatment except in shoot of *C. pennisetiformis*, which had almost equal to its control at 8 and 24 dS/m, respectively. *Puccinellia distans* and *P. divisum* had significantly higher shoot $Cl^-$ than the other species at 8 and 16, and 24 dS/m, respectively. *Panicum turgidum* had the highest root $Cl^-$ concentration of all five species at all salt treatments (Tab. 4).

**Table 4.** Effect of salinity on ion accumulation of some grasses

| Grass species | $Na^+$mmol $kg^{-1}$ fresh weight) | | | | $Cl^-$ (mmol $kg^{-1}$ fresh weight) | | | |
|---|---|---|---|---|---|---|---|---|
| | dS/m | | | | dS/m | | | |
| | Control | 8 | 16 | 24 | Control | 8 | 16 | 24 |
| | Shoot | | | | Shoot | | | |
| *Cenchrus pennisetiformis* | 194.3 ab | 236.4 a | 245.5 a | 228.6 ab | 193.1 a | 186.2 b | 287.7 ab | 227.0 c |
| *Leptochloa fusca* | 213.5 a | 260.5 a | 112.2 b | 287.2 a | 176.7 a | 258.0 b | 238.3 b | 484.0 a |
| *Panicum turgidum* | 248.1 a | 213.9 a | 88.2 b | 194.3 b | 129.8 a | 222.9 b | 249.0 b | 290.4 bc |
| *Pennisetum divisum* | 145.5 c | 225.2 a | 277.3 a | 257.4 a | 148.0 a | 224.6 b | 240.1 b | 598.2 a |
| *Puccinellia distans* | 178.6 b | 105.8 b | 145.9 b | 180.2 b | 136.4 a | 370.4 a | 372.5 a | 349.0 b |
| | Root | | | | Root | | | |
| *Cenchrus pennisetiformis* | 31.8a | 31.8 c | 158.6b | 249.8 b | 30.4 a | 109.6 b | 164.9 b | 126.0 b |
| *Leptochloa fusca* | 34.1 a | 141.6 b | 129.8 b | 169.6 b | 42.9 a | 75.6 c | 82.1 c | 152.6 b |
| *Panicum turgidum* | 48.9 a | 238.3 a | 254.6 a | 503.5 a | 88.0 a | 162.1 a | 217.1 a | 440.4 a |
| *Pennisetum divisum* | 27.4 a | 22.9 c | 156.7 b | 189.6 b | 42.2 a | 83.4 bc | 85.4 c | 185.0 b |
| *Puccinellia distans* | 39.5 a | 32.1c | 109.3 b | 167.5 b | 56.3 a | 55.6 c | 90.2 c | 179.3 b |
| | $Ca^{2+}$ (mmol $kg^{-1}$fresh weight) | | | | $K^+$ (mmol $kg^{-1}$fresh weight) | | | |
| | Shoot | | | | Shoot | | | |
| *Cenchrus pennisetiformis* | 148.4 a | 115.6 a | 194.7 a | 153.2 b | 212.1 a | 136.4 ab | 100.5 a | 108.2 b |
| *Leptochloa fusca* | 144.8 a | 105.2 a | 140.0 b | 219.8 a | 205.5 a | 133.1 ab | 98.6 a | 169.0 a |
| *Panicum turgidum* | 82.4 b | 106.6 a | 114.6 c | 164.9 b | 262.4 a | 152.3 a | 129.1 a | 137.9 ab |
| *Pennisetum divisum* | 101.6 b | 58.7 b | 146.0 b | 118.6 c | 216.7 a | 117.9 ab | 149.2 a | 124.1 ab |
| *Puccinellia distans* | 90.8 b | 90.3 a | 148.2 b | 116.7 c | 137.9 b | 103.4 b | 131.3 a | 94.0 b |
| | Root | | | | Root | | | |
| *Cenchrus pennisetiformis* | 66.5 d | 56.0 c | 114.8 a | 120.2 ab | 32.6 b | 31.7 b | 36.8 c | 29.1 c |
| *Leptochloa fusca* | 89.7 c | 85.2 b | 47.3 b | 133.3 a | 43.4 a | 39.0 ab | 37.0 c | 60.4 b |
| *Panicum turgidum* | 132.2 a | 209.9 a | 123.0 a | 107.4 b | 85.8 ab | 41.6 a | 102.9 a | 136.4 a |
| *Pennisetum divisum* | 104.8 bc | 56.7 c | 51.8 b | 84.2 c | 31.9 ab | 30.0 b | 65.0 b | 13.3 c |
| *Puccinellia distans* | 114.2 b | 76.7 b | 65.8 b | 109.9 b | 51.6 b | 59.4 a | 33.9 c | 60.7 b |

Shoot $K^+$ concentrations (Tab. 4) of all species decreased significantly with increase in salinity level, whereas root $K^+$ of all species remained unaffected. *Pennisetum divisum* and *L. fusca* contained relatively higher shoot $K^+$ at 16 and 24 dS/m, respectively. *Puccinellia distans* showed relatively lower shoot $K^+$ at 8 and 24 dS/m. *Panicum turgidum* had the highest root $K^+$ of all five species at 16 and 24 dS/m.

Shoot $Ca^{2+}$ concentrations of *P. turgidum* and *P. distans* increased with increase in salinity level, whereas the remaining three species did not show any consistent pattern of increase or decrease in shoot $Ca^{2+}$. *Cenchrus pennisetiformis* and *L. fusca* accumulated relatively greater amount of $Ca^{2+}$ in the shoots than the other species at 8 and 16, and 24 dS/m, respectively. Increasing salinity had no significant effect on root $Ca^{2+}$ of all five species. *Panicum turgidum* had significantly greater concentrations of root $Ca^{2+}$ compared with the other species at 8 and 16 dS/m (Tab. 4).

In the present study the salinity treatments were prepared by mixing different salts in ratios that correspond to the composition of subsoil saline water from the deserts. Therefore, considerable interaction of different ions in all species has been observed at each salinity level. Each species used its own specific selective ion transport mechanism in response to varying salinity treatments. For instance, the highly tolerant *Leptochloa fusca* accumulated relatively greater concentrations of $Na^+$ and $Cl^-$ in the shoots at the highest salinity level, whereas these concentrations were low in the roots. Thus, the species used a typical halophytic mechanism [26]. Since the species possesses characteristic salt glands [1] it is possible that $Na^+$ and $Cl^-$ absorbed by roots are rapidly translocated to leaves for onward excretion through salt glands. *Leptochloa fusca* also accumulated high concentrations of both $K^+$ and $Ca^{2+}$ for maintaining $Na^+/K^+$ and $Na^+/Ca^{2+}$ ratios low in the shoots. High $Na^+/K^+$ and $Na^+/Ca^{2+}$ ratios have already been found responsible for increasing membrane permeability in plants [27, 28].

In contrast, the second highly salt tolerant species, *P. distans* partially included $Na^+$ from both shoots and roots, but accumulated high concentration of $Cl^-$ in the shoots. The low concentrations of both $Na^+$ and $K^+$ in the shoots of *P. distans* show that it does not show selectivity to both $K^+$ and $Na^+$, as was suggested by Greenway and Munns [27] that some mesophytes are selective to $K^+$ while others are selective to both $K^+$ and $Na^+$.

The relatively most salt sensitive species, *P. divisum* showed a clear relationship between its poor growth and patterns of ion accumulation under saline conditions. Its high accumulation of both $Na^+$ and $Cl^-$ in the shoots can be related to the early findings of Wyn Jones et al. (1984) [25] who demonstrated that *Agropyron intermedium* was salt sensitive compared with *Agropyron junceum* because it efficiently accumulated both $Na^+$ and $Cl^-$ in its leaves. In addition, low $Na^+/K^+$ and $Na^+/Ca^{2+}$ ratios were not maintained by this species at varying salinity treatments.

The other two species, *C. pennisetiformis* and *P. turgidum* were relatively intermediate in salinity tolerance. *Cenchrus pennisetiformis* absorbed large amount of $Na^+$ in the shoots, but at the same time it accumulated high concentrations of $Ca^{2+}$ in shoot so as to maintain $Na^+/Ca^{2+}$ ratio low. In contrast, *P. turgidum* maintained low concentrations of both $Na^+$ and $Cl^-$ in the shoots, although the concentrations of these ions were high in its roots. This type of mechanism is very common in many salt tolerant mesophytes ([27, 29, 30] Netondo et al., 2004). The same authors advocated that salt excluders have the ability to restrict the uptake of salts into the shoot. This might be due to the phenomenon that toxic ions such as $Na^+$ are absorbed in considerable amount, but are reabsorbed from the root or the shoot and is either stored in the roots or retranslocated to the soil.

Taken overall, it is not difficult to say that *C. pennisetiformis* and *P. turgidum* intermediate in salt tolerance can be grown in those drought hit areas having moderately saline subsoil water for irrigation. These species have high forage value for all types of livestock and already well adapted to the prevailing environmental conditions of the area[10]. Thus, these two species could be of great value for economic utilization of the desert area. The other two highly salt tolerant species *L. fusca* and *P. distans* may not be suitable for the area as they both are adapted to entirely different environmental conditions than those of deserts and also are highly sensitive to drought conditions.

With the proper management of arid lands, the production of these grasses can be enhanced which can bring socio-economic gain of the farmers rehabilitated in these areas. Our present study also confirms that these grasses can be introduced on salt affected and drought prone areas where livestock is facing severe problems of fodder shortage.

## References


1 Malik KA, Aslam Z, Naqvi M (1986) Kallar grass – A plant for saline land. The Neuclear Institute for Agriculture and Biology, Faisalabad, Pakistan
2 Ashraf M, McNeilly T, Bradshaw AD (1987) Selection and heritability of salt tolerance in four forage species. *Crop Sci* 227: 232–243
3 Shannon MC (1985) Principles and strategies in breeding for higher salt tolerance. *Plant Soil* 89: 227–241
4 Ashraf M (2004). Some important physiological selection criteria for salt tolerance in plants. *Flora* 199: 361–376
5 Bates LS, Waldren RP, Tear, ID (1973). Rapid determination of free proline for water stress studies. *Plant Soil* 39: 205–207
6 Witham FH, Blaydes DF, Devlin RM (1971) Experiments in plant physiology. Van Nostrand-Reinhold Co., New York
7 Silva-Fernandes MS, Baker EA, Martin TJ (1964) Studies on plant cuticle. VI. The isolation and fractionation of cuticular waxes. *Ann Appl Biol* 53: 43–58
8 Lowry OH, Rosembrough NJ, Farr AL, Randall RJ (1951) Protein measurement with the Folin phenol reagent. *J Biol Chem* 193(1): 267–275
9 Thomas H (1987) Physiological responses to drought of *Lolium perenne* L. Measurement of, and genetic variation in, water potential, solute potential, elasticity and cell hydration. *J Exp Bot* 38: 115–125
10 Ashraf M, Bokhari MH (1987) Biological approach for economic utilization of the Cholistan desert. *Biologia* 33: 27–34
11 Hsiao TC (1973) Plant responses to water stress. *Annu Rev Plant Physiol* 24: 519–570
12 Quisenberry JE (1982) Breeding for drought resistance and plant water use efficiency. In: MN Christiansen, CF Lewis (eds.): *Breeding Crops for Less Favourable Environments*, John Willey and Sons, New York 193–212
13 Blum A (1989) Osmotic adjustment and growth of barley genotypes under drought stress. *Crop Sci* 29: 230–233
14 Tyree MT, Jarvis PG (1982) Water in tissues and cells. In: OL Lange, PS Nobel, CB Osmond, H Zeiyler (eds.): *Physiological Plant Ecology. II. Water Relations and Carbon Assimilation.* Encyclopedia of Plant Physiology, New Series, Vol. 12B, Springer-Verlag, Berlin 35–37
15 Baker DA (1984) Water relations. In: MB Wilkins (ed.): *Advanced Plant Physiology*. The Pitman Press, Bath 297–318
16 Dainty J (1976) Water relations of plant cells. In: U Lutlge, MG Pitman (eds.): *Transport in Plants II, Part A, Cells: Encyclopedia of Plant Physiology*, New Series, Vol. 2. Springer-Verlag, Berlin 12–35
17 Strogonov BP, Kabanov VV, Shevajakova NJ, Lapine LP, Kamizerko EI, Popove BA, Dostonova RK, Povykod KOLS (1970) *Structure and Function of Plant Cell in Saline Habit.* Nauka Moscow (Trans. Eng.). John Wiley and Sons, New York
18 Bartels D, Nelson D (1994) Approaches to improve stress tolerance using molecular genetics. *Plant Cell Environ* 17: 659–667
19 Hanson AD, Rathinasabapathi B, Rivoal J, Burnet M, Dillon MO, Gage DA (1994) Osmoprotective compounds in the Plumbaginaceae. *Proc Natl Acad Sci USA* 91: 306–310

20 Rudulier DL, Storm AR, Dandekar AM, Smith LT, Valentine RC (1984) Molecular biology of osmoregulation. *Science* 224: 1064–1068
21 Bohnert HJ, Nelson DE, Jensen RG (1995) Adaptation to environmental stresses. *Plant Cell* 7: 1099–1111
22 Ganogopadhyay G, Basu S (2000) Proline as osmotic stress marker in *in vitro* system. In: A Hemantaranjan (ed.): *Plant Physiology, Biochemistry and Plant Molecular Biology in 2000*. Vol. 3. Scientific Publishers, Jodhpur-India 283–304
23 Jordan WR, Shouse PJ, Blum A, Miller FR, Monk PK (1984) Environmental Physiology of Sorghum. II. Epicuticular wax load and cuticular transpiration. *Crop Sci* 24: 1168–1173
24 Ashraf M, Mehmood S (1990) Responses of four *Brassica* species to drought stress. *Environ Exp Bot* 30: 93–100
25 Wyn Jones RG, Gorham J, McDonnell E (1984) Organic and nonorganic solute contents as selection criteria for salt tolerance in the Triticeae. In: RC Staples, GH Toenniessen (eds.): *Salinity tolerance in plants-Strategies for crop improvement*, John Wiley and Sons, New York, 189–203
26 Flowers TJ, Troke PF, Yeo AR (1977) The mechanism of salt tolerance in halophytes. *Annu Rev Plant Physiol* 28: 89–121
27 Greenway H, Munns RA (1980) Mechanism of salt tolerance in non-halophytes. *Ann Rev Plant Physiol* 31: 149–190
28 Muhammad S, Akbar M, Neue HU (1987) Effect of Na/Ca and Na/K ratios in saline culture solution on the growth and mineral nutrition of rice (*Oryza sativa* L.). *Plant Soil* 104: 57–62
29 Qain YL, Wilhelm SJ, Marcum KB (2001) Comparative responses of two Kentucky bluegrass cultivars to salinity stress. *Crop Sci* 41: 1895–1900
30 Netondo GW, Onyango JC, Beck E (2004) Sorghum and salinity. II. Gas exchange and chlorophyll florescence of sorghum under salt stress. *Crop Sci* 44: 806–811

Biosaline Agriculture and Salinity Tolerance in Plants
Edited by M. Öztürk, Y. Waisel, M.A. Khan and G. Görk


# Salt and seawater effects on the germination of *Crithmum maritimum*

Abdallah Atia, Karim Ben Hamed, Ahmed Debez and Chedly Abdelly

*Laboratoire d'Adaptation des Plantes aux Stress Abiotiques, INRST, BP 95, Hammam-Lif, 2050, Tunisia*

## Introduction

*Crithmum maritimum* (*Apiaceae*), also called sea fennel, is a perennial halophyte that thrives on saline environments (rocky coasts, piers and breakwaters) along the Mediterranean countries, Pacific and Atlantic coasts [1]. Several uses are known for this plant: for culinary purposes, fresh leaves and young branches are pickled in vinegar and used as condiments. Leaves have also medicinal applications, as antiscorbutic, tonic, diuretic, and vermifuge substances [2].

Salt tolerance during the germination step is critical for the successful establishment of the future plant. Vascular species often show a delay and a reduction in germination as salt levels in the medium increase [3]. The reduction in germination may be due to the low osmotic potentials occurring under saline conditions or to the toxic effects of the ions. Nevertheless, an important characteristic that distinguishes halophytes from glycophytes is their ability to remain dormant at high salinities and to germinate later when the level of salinity is reduced.

We studied in the present work the germination response of *C. maritimum* seeds to NaCl and sea water, as major salinity agents, in order to identify the threshold salinity for a significant reduction in germination.

## Material and methods

Seeds of *C. maritimum* were collected from Tabarka (North West of Tunisia) and kept dry in a cold room at 4° C. Sterilised seeds were germinated in Petri dishes, on filter paper moistened with either distilled water or saline solutions containing either NaCl (50, 100, 150, 200, 300 mM) or sea water at different dilutions (5, 10, 20, 30 %). Four replicates of 25 seeds each were used for each treatment. Seeds were considered to germinate at the emergence of the radicle. Seed germination was recorded each 2 days for 46 days. Seeds which failed to germinate in saline solutions were transferred in distilled water to assess seed viability.

Germination rate was calculated using a modified Timson index of germination velocity; $\sum G/t$, where $G$ is the percentage of seed germination after 2 d intervals, and $t$ is the total time of germination [4]. Rate of recovery of germination was calculated using the relation $[(a - b)/(c - b)]$ 100, where $a$ is the total number of seeds germinated after beg transferred to distilled water, $b$ is the total number of seeds germinated in saline solution, and $c$ is the total number of seeds.

A one-way variance analysis (ANOVA) was carried out by the statistical software SPSS 10.0.

## Results

The evolution of the germination percentage of *C. maritimum* over the salt-treatment period showed that the germination process was strongly inhibited when sea water dilution exceeded 5 % (Fig. 1). Sea water at 5 % delayed the germination but enhanced the germination plateau up to 85 % after 20 days, while reaching 75 % in distilled water. The variation of the germination capacity of *C. maritimum* in the presence of NaCl showed that the germination was delayed under increasing NaCl, and strongly inhibited at NaCl concentrations higher than 50 mM. Oneway ANOVA revealed that the final germination percentage was significantly inhibited by both sea water and NaCl (Figs. 2A and 2B). The same pattern appeared when considering the germination rate (Figs. 2C and 2D), thus confirming the salt-induced germination delay.

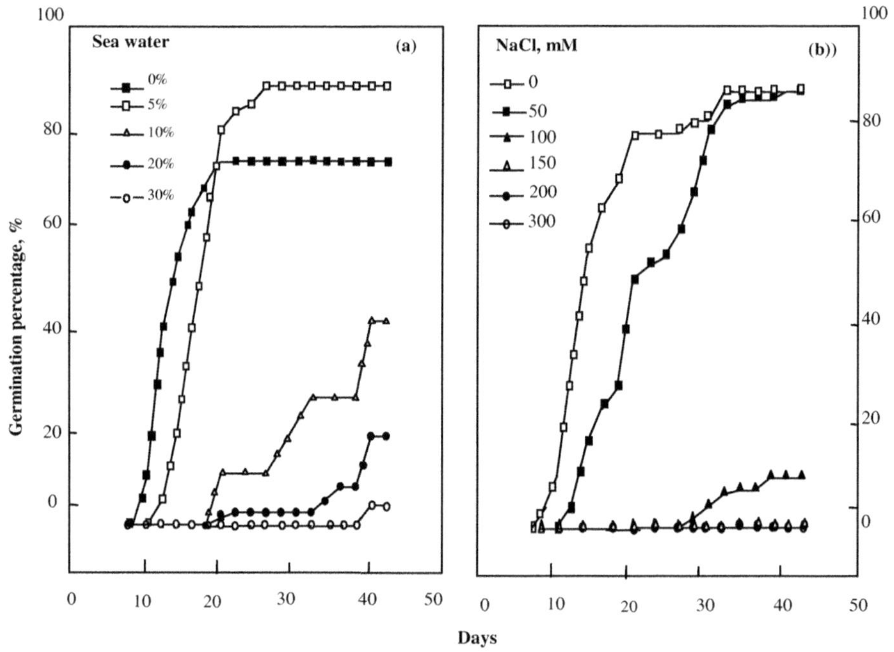


**Figure 1.** Effect of salinity (**a**: sea water, **b**: NaCl) on the germination (%) of *Crithmum maritimum*

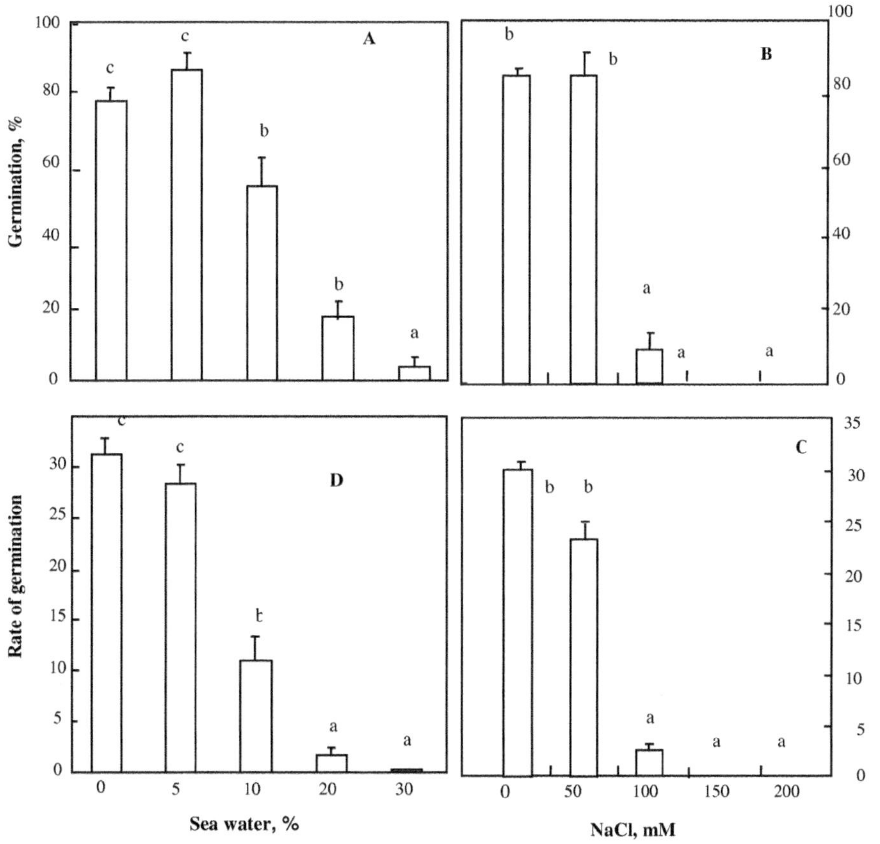


**Figure 2.** Effect of salinity on germination parameters of *C. maritimum*. **A** and **B**: mean final germination percentage under sea water and NaCl, respectively. **C** and **D**: Rate of germination (Timson's germination velocity index) under sea water and NaCl, respectively. The two parameters were determined after 46 days of treatment. Different letters indicate statistically significant differences ($P < 0.05$) between treatments (by Tukey's test).

The inhibition of germination could be due either to salt-induced seed mortality, or to unfavourable external osmotic conditions. In order to distinguish between these two factors, the seeds which did not germinate in the presence of salt were transferred on pure water. Nine days later, up to 90 % germination recovery was observed (Fig. 3), which suggests that NaCl did not alter seed viability. Thus, the observed inhibition of germination following NaCl treatment seems to be mainly attributable to unfavourable osmotic conditions, although NaCl toxic effect was observable at the highest concentrations.

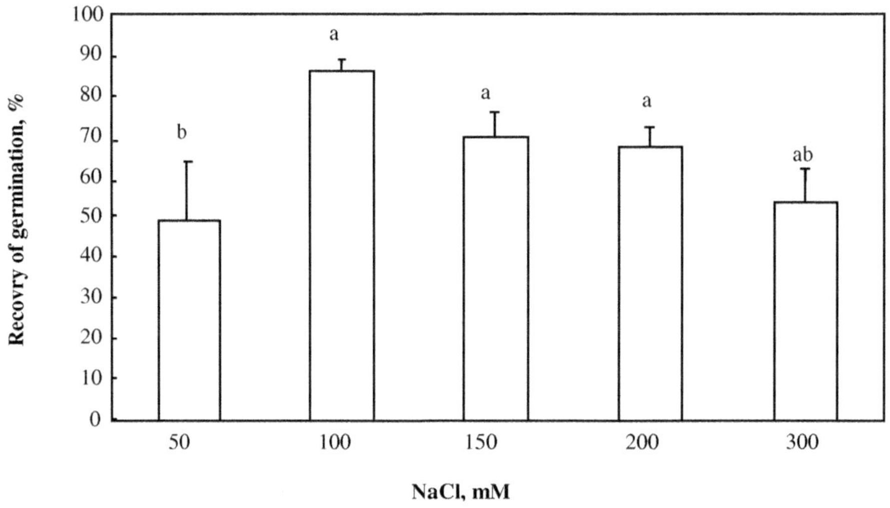


**Figure 3.** Germination recovery (%) of *C. maritimum*. All seeds which did not germinate after 46 days in the presence of 50, 100, 150, 200, and 300 mM NaCl, were germinated in pure water and recovery of germination capacity was determined as indicated in *Material and methods*. Different letters indicate statistically significant differences ($P < 0.05$) between treatments (by Tukey's test).

## Discussion

Seed germination was delayed in the presence of NaCl and seawater and inhibited when NaCl concentration exceeded 100 mM and 5 % seawater dilution. Similar trends have been reported in other halophytes, which generally germinated better under reduced non-saline conditions. For instance, the germination of *Urochondra setulosa* was maximal in distilled water (100 %), but decreased to 20 % at 300 mM NaCl [5]. It has been reported that germination was substantially inhibited at 20 $g.l^{-1}$ NaCl in *Atriplex patula* [6]. Despite the depressive effect of salinity, halophytes are able to germinate at salt concentrations similar to that of sea water or even higher: in this way, seeds of *Triglochin maritima* germinated even at 400 mM NaCl [7], while *Kochia scoparia* germinated with a rate of 30 % at 1,000 mM NaCl [8]. The germination aptitude of *Spartina alterniflora* was preserved even at 1,027 mM, even with a very low rate of 8 % [9]. In addition, it was reported that the seeds of *Kochia americana* are able to germinate even at 1,712 mM NaCl [8].

Our results indicate that salt inhibited germination without damaging the seeds, which could recover their capacity to germinate when transferred to pure water. Hence, the main factor involved in the salt induced dormancy of *C. maritimum* seems to be the low water potential of the medium, as described for *Carpobrotus* [10], *Suaeda fruticosa*, *Triglochin maritima* [7] and *Aeluropus lagopoides* [11]. Our data support the assumption that salinity usually impairs germination by preventing imbibition [12].

The fact that salt pre-treatment did not affect seed viability could have a great significance for the ecophysiology of *C. maritimum*. Indeed, this finding suggests that *C. maritimum* would be able to constitute viable seeds bank when salinity levels are high, and that these seeds would germinate early in the spring, after salt leaching from the soil surface by the winter precipitations. This hypothesis should be checked with long salt treatments of seeds, to document the mortality kinetics of seeds in various conditions of salt and humidity.

## References

1 Tutin TG, Heywood VH, Burges NA, More DM, Valentines DH, Walters SM, Webb DA (eds.) (1968) *Flora Europea* Vol. 2, Cambridge University Press, Cambridge
2 Özcan M (2000) The use of yogurt as starter in rock samphire (*Crithmum maritimum* L.) fermentation. *Eur Food Res Technol* 210: 424–426
3 Ungar IA (1995) Seed germination and seed-bank ecology in halophytes. In: *Seed Development and seed Germination*. Marcel Dekker, New York 599–628
4 Khan MA, Ungar IA (1984) Effects of salinity and temperature on the germination and growth of *Atriplex triangularis* Willd. *Am J Bot* 71: 481–489
5 Gulzar S, Khan MA, Ungar IA (2001) Effect of temperature and salinity on the germination of *Urochondra setulosa*. *Seed Sci Res* 29: 21–29
6 Ungar IA (1996) Effects of salinity on seed germination, growth and ion accumulation of *Atriplex patula*. *Am J Bot* 83: 604–607
7 Khan MA, Ungar IA (1997) Effect of light, salinity, and thermoperiod on the germination of halophytes. *Can J Bot* 75: 835–841
8 Khan MA, Gul B, Weber DJ (2001) Effect of salinity and temperature on the germination of *Kochia scoparia*. *Wetl Ecol Manage* 9: 483–489
9 Moorning MT, Cooper AW, Seneca ED (1971) Seed germination response and evidence for height eciphenes in *Spartina Alterniflora* from North Carolina. *Am J Bot* 58: 48–55
10 Weber E, d'Antonio C (1999) Germination and growth responses of hybridising *Carpobrotus* species (Aizoaceae) from coastal California to soil salinity. *Am J Bot* 86: 1257–1263
11 Gulzar S, Khan MA (2001) Seed germination of a halophytic grass *Aeluropus lagopoides*. *Ann Bot* 87: 3119–3324
12 Keiffer CH, Ungar IA (1995) Germination responses of halophytes exposed to prolonged hypersaline conditions. In: MA Khan, IA Ungar (eds.): *Biology of salt tolerant plants*. Department of Botany, University of Karachi, Karachi, Pakistan 43–50

Biosaline Agriculture and Salinity Tolerance in Plants
Edited by M. Öztürk, Y. Waisel, M.A. Khan and G. Görk


# Growth performance and nutritional value of salt tolerant plants growing under saline environments

Muhammad Y. Ashraf[1], Ghulam Sarwar[1], Muhammad Ashraf[2], Faqir Hussain[1], Riaz A. Wahed[1] and Muhammad M. Iqbal[1]

[1] *Stress Physiology and Plant Biochemistry Lab, Salinity and Environmental Management Division, Nuclear Institute for Agriculture and Biology, P.O. Box 128, Jhang Road, Faisalabad, Pakistan*
[2] *Department of Botany, University of Agriculture, Faisalabad (38040), Pakistan*

## Socio-economic effects of salinity

Salinity is a serious problem of agriculture worldwide, particularly in arid and semi-arid regions. Most of Pakistan is arid to semi-arid with low annual precipitation and 6.3 million hectares (mha) land is affected to varying degree of salinity [1]. As a result of this, heavy losses in crop yields and plant productivity have been recorded [2]. Lands with high salinity are not cultivated and are changing into wasteland, and farmers owning these lands are migrating towards cities or towns as a result of which population load on cities is increasing day by day and deficiencies in food, feed, fodder and industrial materials are being faced. So, there is an urgent need to utilize these lands for plant production [3].

## Halophytes as livestock fodder

Halophytes having economic values can be grown on salt-affected wastelands, which may bring socio-economic gains. It was observed that farmers as well as livestock of salt-affected areas face great scarcity of food, fodder and feed. The farmers of these areas often arrange fodders for their livestock from other areas at very high costs. However, if these areas are utilized for developing pastures or rangelands with salt tolerant forage plants, good economic returns can be achieved. In Pakistan, Nuclear Institute for Agriculture and Biology (NIAB), Faisalabad, Pakistan, started work in the 1970s to identify the salt tolerant plants with economic values so that they may be utilized as food, fodder and industrial raw materials. Dozens of salt tolerant plants with economical value have been identified and introduced at two biosaline research stations. Some of these can be utilized as fodder; however, farmers are reluctant to use them as fodder due to their high salt content. It is necessary to have complete knowledge about the nutritive values of these plants, so that they can be recommended to the farmers for using them as fodder for their livestock.

The studies have been conducted at NIAB, Faisalabad, Pakistan, with the aim to work out the forage values and their chemical composition. On the basis of their utility as forage, these can be recommended for the livestock in saline areas and thereby salt-affected wastelands can be utilized for better economic returns. So, the present study was conducted with five halophyte forage plants, two were grasses and the other three were shrubs, i.e., *Leptochloa fusca*, *Sporobolus arabicus*, *Suaeda fruticosa*, *Kochia indica* and *Atriplex lentiformis*, respectively.

## Cultivation of halophyte grasses, bushes and their chemical analysis

Studies were conducted at the Biosaline Research Station-II (BSRS-II) of Nuclear Institute for Agriculture and Biology (NIAB), Faisalabad, Pakistan, situated 40 km from Faisalabad, Pakistan. The salt concentration in ground water is 4,000–6,000 ppm and the water table varies from 2.1–3.0 m, while soil salinity ranges from 12–27 dS $m^{-1}$. The other characteristics of soil and water are summarized in Table 1. The site has an annual average temperature of 32°C. The annual average precipitation and evaporation is 320 mm and 1,100 mm, respectively. The halophyte species viz, *Leptochloa fusca* (Kallar grass), (*Sporobolus* grass), *Suaeda fruticosa* (Lana), *Kochia indica* (Kochia) and *Atriplex lentiformis* (salt bush) were grown from stubbles and seeds in plots of 10 × 10 m, repeated four times; side by side naturally growing plants of these species were also selected. The irrigation water was underground brackish water with high salt (the characteristics of tubewell water are summarized in Table 2). When the plants were at the feeding stage, samples were collected from cultivated plots and naturally growing plants at BSRS-II. Their fresh and dry weights were recorded and to evaluate their nutritive values, samples were oven dried. The leaves and twigs were ground and used for subsequent chemical analyses. Crude protein (Nx6.25), fibers, total minerals and total carbohydrate were determined according to AOAC [4] procedures. The above ground plant material was digested according to

**Table 1.** Characteristics of soil of study site, BSRS-II Pacca Anna, NIAB, Faisalabad, Pakistan

| **Soil characteristics** | **Range** |
|---|---|
| Soil texture | Sandy loam |
| Clay (%) | 14 ± 1.4 |
| Silt (%) | 18 ± 1.5 |
| Sand (%) | 68 ± 3.1 |
| EC (dS $m^{-1}$) | 12–27.24 |
| Saturation Percentage | 25.36–31.45 |
| PH | 7.82–8.92 |
| Bulk density (g $cm^{-3}$) | 1.38–1.58 |
| $CaCO_3$ (%) | 12–23 |
| $CaSO_4.2H_2O$ (%) | 2.56–4.15 |

**Table 2.** Characteristics of Tube-well water at BSRS-II Pacca Anna, NIAB, Faisalabad, Pakistan

| Characteristics | Range |
|---|---|
| EC (dS $m^{-1}$) | 4.97 |
| PH | 8.2 |
| TSS | 3878 |
| SAR | 40.5 |
| SAR adj | 101.25 |
| RSC | 21.60 |
| Soluble ions (me $L^{-1}$) | |
| $Na^+$ | 51.2 |
| $K^+$ | 0.4 |
| $Ca^{2+} + Mg^{2+}$ | 3.21 |
| $Cl^-$ | 13.75 |
| $CO_3$ | 1.5 |
| $HCO_3$ | 21.75 |
| $SO_4$ | 17.35 |

Wolf [5] and cations like $Na^+$, $K^+$ and $Ca^{2+}$ were estimated by flame photometer (FP7, Jenway, England) and from the same aliquot Mg was determined titrimetrically as described in US Salinity Lab Hand Book-60 and P using Barton's reagent [6]. The data were statistically analyzed [7].

## Growth and yield of halophyte plants

The yield per plot is not shown because it is obvious that the plant with higher salt tolerance potential had higher yield and that was maximum in *Atriplex*, closely followed by *Suaeda fruticosa* and *Kochia indica* and minimum was in *Sporobolus*. The yield was higher in experimental plots than that of plants growing naturally. The pattern was almost similar in dry matter accumulation (Fig. 1a). As *Leptochloa fusca* and *Sporobolus grasses* excrete salt absorbed through the salt glands present in their leaves [8], so these species contained more water than other salt tolerant bushes [9]. The salt tolerant bushes contained more salts and less water, as a result they had more dry matter and total minerals. The palatability reports indicated that *Atriplex* and other salt bushes are not as palatable as *Sporobolus* and *Leptochloa fusca* and other grasses [9–11]. Similarly, Casson et al. [12] reported that higher dry matter in salt tolerant plants is due to high accumulation of salts, which is very clear from the present study, because the total mineral contents were very high in the salt tolerant bushes.

## Mineral and nutrient values of halophytes

The total minerals were the maximum in *Atriplex lentiformis*, followed by *Suaeda fruticosa*, *Kochia indica*, *Leptochloa fusca* and *Sporo-bolus arabicus* (Fig. 1b), which

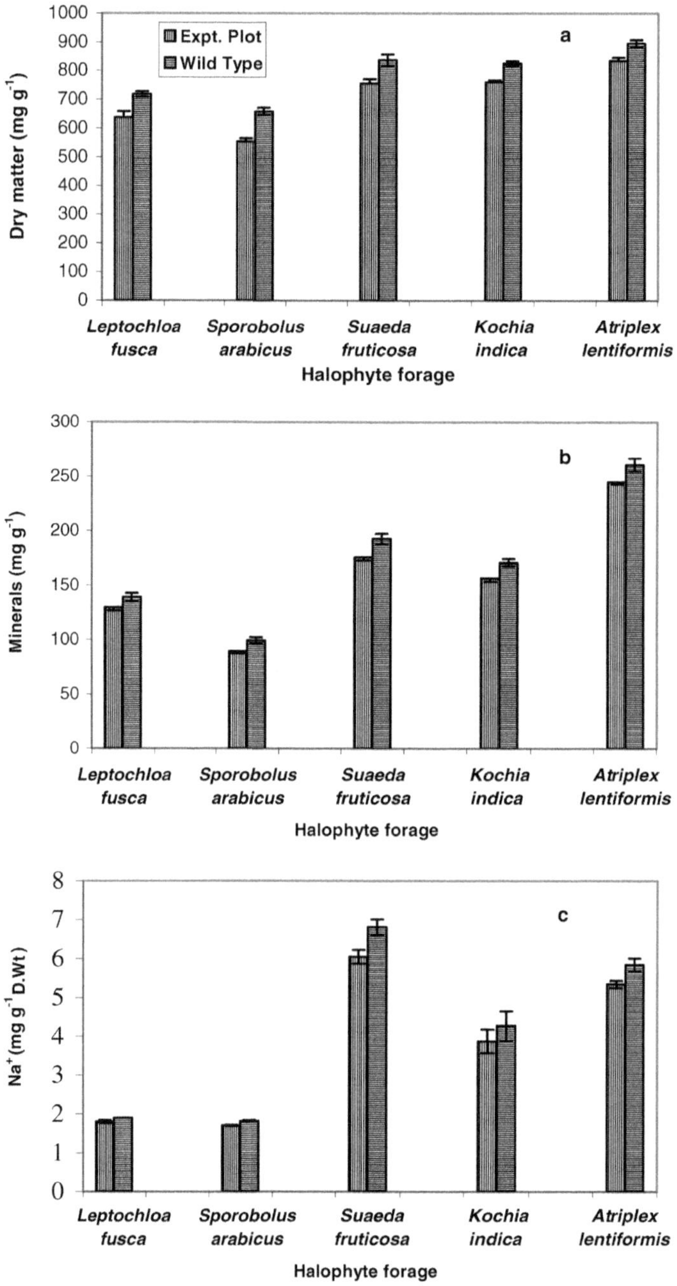


**Figure 1.** Growth performance and nutritional value of some halophyte forage species under saline environments (**a**) dry weight (**b**) total minerals (**c**) sodium contents

confirmed that maximum salts were accumulated in *Atriplex* and minimum in *Sporobolus*. On the basis of salt concentration, *Sporobolus* is categorized as the most palatable grass for livestock. Many reports indicated that the plants containing higher concentrations of salts are toxic for livestock and are responsible for different types of diseases and physiological disorders. The literature indicates that the animal fed on plants having higher concentrations of salts cause lesion and rashes in the stomach of the animals [13].

Chemical composition of the plants for ions like $Na^+$ (Fig. 1c), $K^+$ (Fig. 2a), $Ca^{2+}$ (Fig. 2b) and $Mg^{2+}$ (Fig. 2c) also varied significantly within different halophytic forage species. Maximum $Na^+$ was recorded in *Suaeda fruticosa* and minimum in *Sporobolus* and *Leptochloa fusca* (Fig. 1c). The literature has clearly indicated that the plants with higher concentrations of $Na^+$ are injurious for animal health and only the *Sporobolus* and *Leptochloa fusca* in experiment plots maintained $Na^+$ concentration which is near to critical limit for livestock. However, the naturally growing plants contained higher $Na^+$ concentrations near the critical limit necessary for livestock (Fig. 1c). So, *Sporobolus* and *Leptochloa* are more suitable as animal fodder than other salt tolerant plants. Many reports [9, 14] have suggested mixing of *Atriplex*, *Suaeda fruticosa* and *Kochia* in animal ration as a result of which $Na^+$ as well as other salt concentrations can be reduced in animal ration and can be made more palatable and digestible.

Potassium concentrations were higher in *Kochia*, closely followed by *Atriplex* and *Suaeda fruticosa*, while it was minimum in *Leptochloa* and *Sporobolus*. Although animals require higher $K^+$ concentration to maintain their optimum metabolic activities, in higher amounts it is toxic (Fig. 2a). The salt tolerant bushes contained higher amounts of $K^+$ than the critical limit; however, salt tolerant grasses contained $K^+$ up to the requirements of the animals. So these grasses can be recommended for direct grazing or feeding. However, for salt bushes, mixing with other plants is necessary.

Salt bushes also maintained relatively higher $Ca^{2+}$ and $Mg^{2+}$. However, maximum $Ca^{2+}$ was in *Atriplex*. *Suaeda fruticosa* contained the highest $Mg^{2+}$ concentration (Figs. 2b and c). The grasses however, contained low amounts of these salts. This may be due to the excretion of salts through salt glands, which are absent in salt bushes [8]. The earlier work of NIAB scientists has demonstrated that these grasses have salt glands in their leaves to get rid of the excessive salt concentrations. This property of these plants therefore makes them more palatable and digestible. These grasses are already recommended to farmers for cultivation on salt-affected wasteland for livestock grazing. *Atriplex* and other salt bushes are also being utilized in the animal ration by mixing with other feed. Their sole use is not useful for livestock. Observations indicated that if the animals were only fed on salt bushes, they become weaker day by day and suffer from various metabolic or physiological disorders.

Maximum phosphorous was accumulated in *Suaeda fruticosa* and the minimum in *Sporobolus,* closely followed by *Leptochloa*. The salt bushes again contained more P than the critical limit for the livestock (Fig. 3a). P is a nutrient required in higher amounts for animals but its excessive amount may disturb the other metabolic functions of livestock. Reports have indicated [15, 16] that the salt grasses are useful for animals because they contain desirable amounts of P (Fig. 3a).

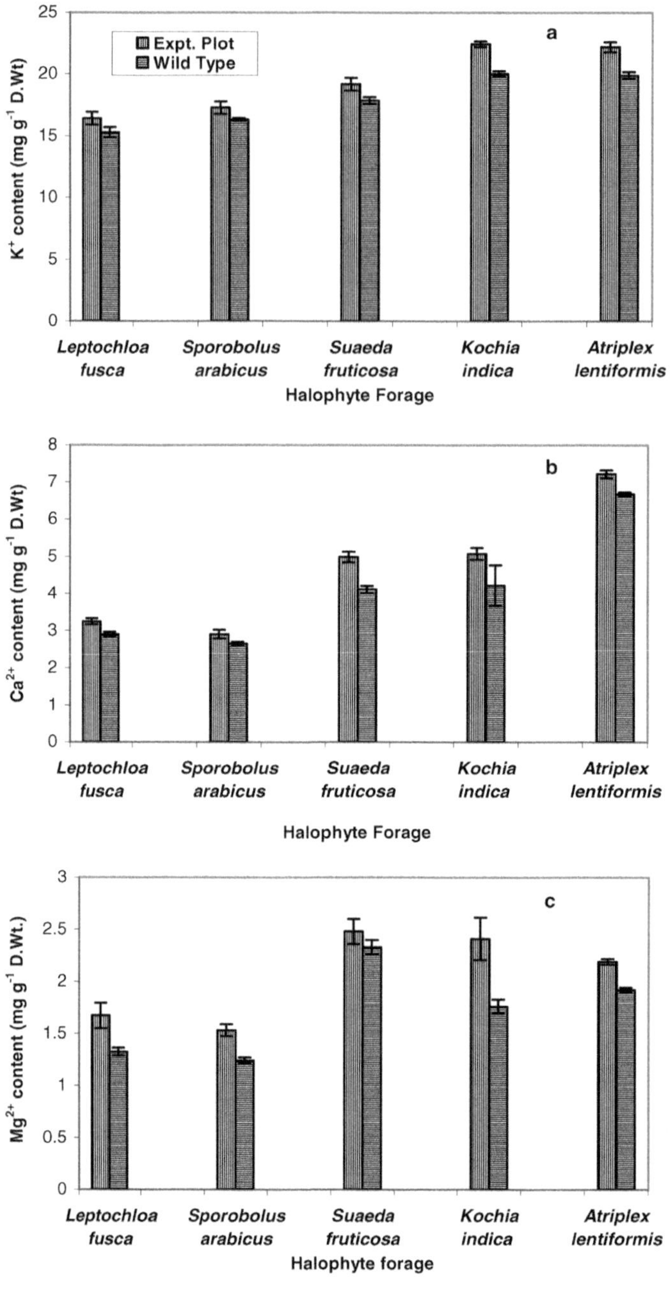


**Figure 2.** Nutritional value of some salt tolerant plants growing under saline environments (**a**) Potassium, (**b**) Calcium (**c**) Magnesium contents

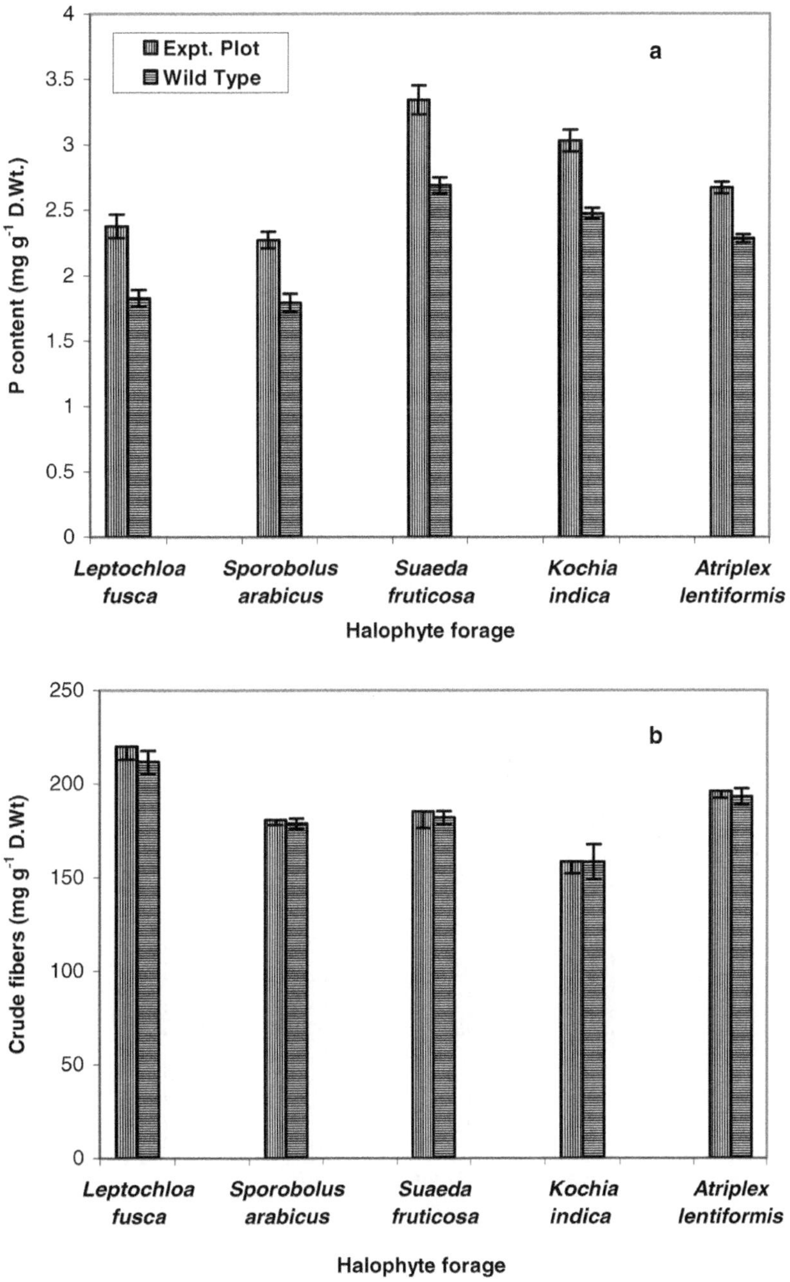


**Figure 3.** Nutritional value of some halophyte forage growing under saline environments (**a**) Phosphorus (**b**) Crude fiber contents

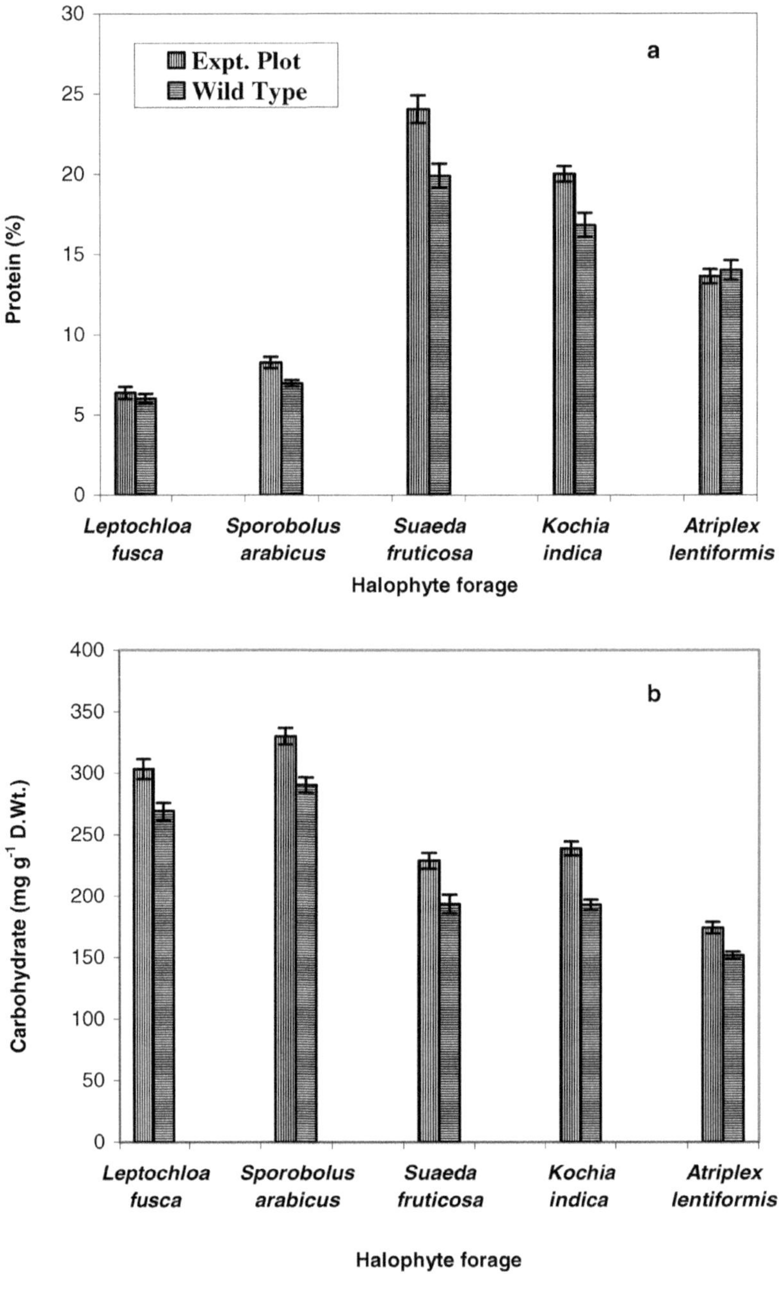


**Figure 4.** Nutritional value of some salt tolerant plants growing under saline environments (**a**) Protein (**b**) Carbohydrate contents

The results regarding *protein* contents of plants (Fig. 4a) also indicated that salt bushes had more protein than salt tolerant grasses, while reverse was the case for crude fibers (Fig. 3b). As regards the carbohydrates, they were the highest in *Sporobolus arabicus,* closely followed by *Leptochloa fusca* and minimum in *Atriplex lentiformis* (Fig. 4b). These findings also proved that *Sporobolus* and *Leptochloa* grasses are useful for animal health and contain nutrients up to the limit required by the livestock, while salt bushes contained more salts, which could create other metabolic disorders.

## Recommendations

From the findings of the present investigation, it was also very clear that the cultivated plants contained less amounts of salts than ones growing naturally. This may be due to proper irrigation as a result of which they were able to contain reasonable amounts of water that caused dilution of salts, and plants became more suitable as animal fodder.

The palatability checked at NIAB [17] and at other places of the world also indicated that the *Atriplex*, *Suaeda fruticosa* and *Kochia indica* are not as suitable as the *Sporobolus* and *Leptochloa fusca*. The Animal Section of the Biological Chemistry Division of this institute has also conducted the digestibility studies of these species using the Nylon bag technique. It indicated that salt tolerant grasses had more digestibility values than salt bushes. So, on the basis of the above results *Sporobolus arabicus* and *Leptochloa fusca* are more useful and seem close to the normal fodder for livestock. The findings of the Animal Section of the Biological Chemistry Division of this institute has also showed that feeding of *Leptochloa* or *Sporobolus* grass had no adverse effects on health and reproduction of dwarf goats. Thus, they should be cultivated preferably on salt-affected wastelands to produce large amounts of fodder for livestock. The salt tolerant bushes should also be grown, but should be fed by mixing with other fodders containing lesser amount of salts.

## References


1 Anonymous (2004) Statistical Survey of Pakistan 2003–04. Government of Pakistan, Finance Division, Economic Advisor Wing, Islamabad

2 Ashraf MY, Sarwar G (2002) Salt tolerance potential in some members of Brassicaceae: Plysiological studies on water relations and mineral contents. In: R Ahmad, KA Malik (eds.) *Prospects for Saline Agriculture.* Kluwer Academic Publishers, The Netherlands 237–245

3 Aslam Z, Bhatti AS, Mujtaba M (1999) *Aster tripolium*: A winter forage: Introduction for salt-affected soil. In: H Hamdy, H Lieth, M Todorovic, M Moschenko (eds.) *Halophytes Uses in Different Climates – II.* Backhuys Publishers, Leaden, The Netherlands 87–93

4 AOAC (Association of Official Analytical Chemist) (1990) Official Methods of Analysis, $13^{th}$ Edn. AOAC, Washington, D.C.

5 Wolf B (1982) A comprehensive system of leaf analysis and its use for diagnosing crop nutrient status. *Comm Soil Science and Plant Analysis* 13: 1035–1059

6 Jackson ML (1962) Soil Chemical Analysis. Constable and Company Ltd., UK

7 Steel RGD, Torrie JH (1980) Principal and procedures of statistics: A Biometrical Approach 2nd ed., McGraw-Hill, New York
8 Wieneke J, Sarwar G, Roeb M (1987) Existence of salt glands on leaves of Kallar grass (*Leptochloa fusca* L. Kunth). *J Plant Nut* 10: 805–820
9 Aganga AA, Mthetho JK, Tshwenyane S (2003) *Atriplex nummularia* (Old Man Saltbush): A potential forage crop for arid regions of Botswana. *Pak J Nutrition* 2: 72–75
10 Squires VR, Ayoub AL (1992) Halophytes as a resource for livestock and rehabilitation of degraded lands. *Proceeding of the International Workshop on Halophytes for Reclamation of Saline Wastelands and as a resource for livestock problems and prospects.* Nairobi, Kenya. Kluwer Academic Publishers 22–27
11 El Shaer HM, Fahmy AA, Abdul-Aziz GM, Shalaby AS, Abd El Gawd AM (2000) Utilization of less and unpalatable halophytes as non-conventional feed for sheep under the arid condition in Egypt. Proceeding Of the 3rd All Africa Conference of Egyptian Society of Animal Production, Alenndra, Egypt
12 Casson T, Warren BE, Schleuter K, Parker K (1996) On farm sheep production from sheep pastures. *Proc Aust Soc Anim Prod* 21: 173–176
13 Atiq-ur-Rehman, Mackintosh JB, Fortune JA, Warren BE (1994) Can the voluntary feed intake of wheat straw in sheep improved by mixing with saltbush pastures? *Proc Aust Soc Anim Prod* 20: 175–177
14 Ben Salem H, Nefazaoui A (2000) Growth performance of sheep fed on mixed diets containing shrubs (*Acacia, Atriplex, Cactus*) and nitrogen/energy supplementation. In: Gintzburger G, Bounejmate M, Agola C, Mossi K (eds.). Production and utilization of multi-purpose fodder shrubs and trees in West Asia, North Africa and the Sahel. ICARDA, Aleppo, Syria; ILRI, Narobi, Kenya. (ISBN 92-9217-095-4)
15 Venuto BC, Burson BL, Hussey MA, Redfearn DD, Wyatt WE, Brown LP (2003) Forage yield, nutritive value, and grazing tolerance of Dallisgrass biotypes. *Crop Sci* 43: 295–301
16 Burson BL, Watson JP (1995) Bahgrass, dallisgrass and other *Paspalun* species. In: RF Barnes (ed.) An introduction to Grassland Agriculture, Iowa State University Press, Ames, IA 431–440
17 NIAB (Nuclear Institute for Agriculture and Biology) (1997) Economic utilization of salt-affected soils: Twenty Five Years of NIAB. Report on Achievements (Silver Jubilee of NIAB), Publisher NIAB, Faisalabad, Pakistan 123–151 (ISBN-969-8038-09-4)

Biosaline Agriculture and Salinity Tolerance in Plants
Edited by M. Öztürk, Y. Waisel, M.A. Khan and G. Görk


# Comparative effect of NaCl and seawater on seed germination of *Suaeda salsa* and *Atriplex centralasiatica*

Xiaojing Liu[1], Hailong Qiao[1], Weiqiang Li[1], Toshiaki Tadano[2] and M. Ajmal Khan[3]

[1] *Center for Agricultural Resources Research, Institute of Genetics and Developmental Biology, Chinese Academy of Sciences, 286 Huaizhong Road, Shijiazhuang, Hebei 050021, P.R.China*
[2] *Department of Applied Biology and Chemistry, Faculty of Applied Bio-Science, Tokyo University of Agriculture, Tokyo, Japan*
[3] *Department of Botany, University of Karachi, Karachi, Pakistan*

## Introduction

Seed germination in annual halophytes usually occurs when soil salinity levels are low and soil moisture is relatively high [1]. Optimum germination of halophyte seeds is often obtained under freshwater and inhibited by increasing salinity concentrations [2–4], but the ability to germinate at higher salinities is varied with species, for example *Salicornia herbacea* germinated at 1,700 mM NaCl [5], *Arthrocnemum macrostachyum* can germinate at 1,000 mM NaCl solution with 10 % germination percentage [6]. Some secreting halophytes could also germinate above seawater salinity [7–11]. Most secreting halophytes show germination at NaCl concentrations ranging from 0.34–0.52 M NaCl. Few of them have low salt tolerance during germination [12–14].

Seed of halophytes under natural conditions are subjected to saline stress dominated by NaCl. However, other chloride, sulfate and carbonate salts and their interaction play a significant role in affecting seed germination [1, 4]. There has been much research on halophyte seed germination in NaCl solutions. Such tests may not be relevant to the field conditions because in the field the soil solutions contains different cations ($Na^+$, $Mg^{2+}$, $Ca^{2+}$, $K^+$) and anions ($Cl^-$, ${SO_4}^{2-}$, $HCO_3^-$ etc.), which compositions are similar to seawater. Different salt sources have different effects on seed germination [15]. Zia and Khan [16] reported that seawater inhibited seed germination of *Limonium stocksii* more than NaCl solutions, similar to Joshi et al. [17] reported for *Salvadora persica*. However, Duan et al. found that *Suaeda salsa* [18] and *Chenopodium glaucum* [19] germinated better in soil extracted solutions than in NaCl solutions, so did some combinations of salts have the similar effects on *Securigera securidaca* [20] and *Rhus chinensis* [21]. Those compared studies are still limited on the relative tolerance of seawater and NaCl solution on seed germination of halophytes.

Recovery of germination responses in temperate halophytes has been demonstrated in *Salicornia europaea, Suaeda linearis* [22], *Spergularia marina* [9], *Suaeda depressa* [23], *Arthrocnemum australsicum, Triglochin striata, Suaeda australis, Juncus maritimus,* and *Casuarina glauca* [24]. Boorman [25, 26] and Woodell [7] also reported salt stimulation of seed germination following treatment with seawater for a number of saltmarsh species. Keiffer and Ungar [27] exposed seeds of five halophytes (*Atriplex prostrata, Hordeum jubatum, Salicornia europaea, Spergularia marina* and *Suaeda calceoliformis*) to an extended period of salinity treatments and determined their recovery responses when transferred to distilled water. They used Woodell [7] classification system and placed *Atriplex prostrata* seeds in the Type 1, *Hordeum jubatum* and *Spergularia marina* in the Type 2, and *Salicornia europaea* and *Suaeda calceoliformis* in the Type 3 category.

*Suaeda salsa*, a leaf succulent annual plant, and *Atriplex centralasiatica*, a secreting annual plant from the family Chenopodiaceae, are widely distributed in China and are quite common in coastal areas. Both species have the potential to be used as cash crops [28–30]. The present study investigates the effects of seawater and NaCl on the seeds germination of *S. salsa* and *A. centralasiatica* and compared the differential response of these two species.

## Materials and methods

Seeds of *S. salsa* and *A. centralasiatica* were collected in October 2004 in the coastal saline soils of Haixing County, Hebei Province of China. Before storage they were surface sterilized using clorox (0.5 %) for 1 minute followed by thorough rinsing with distilled water and air-drying for a few days. After cleaning, seeds were stored in paper bags at room temperature with relative humidity 30–40 % in the laboratory and germination experiments were initiated in January 2005. Germination experiments were carried out in 10 cm diameter tight fitting plastic Petri dishes with two-layer wet filter paper. 10 mL of test solution were added in the Petri dishes to investigate the effect of salinity on seeds germination of *S. salsa* and *A. centralasiatica*. Germination was carried out in different dilutions (0, 5, 10, 20, 30, 40, 50 dS $m^{-1}$) of NaCl and seawater separately. The electrical conductivity for both salt solutions was maintained at the desired level with the help of a conductivity meter. Four replicates of 40 seeds each were used for all treatments. Seeds with visible radicle were considered to have germinated.

A 24 h cycle was used where day temperature (25° C) coincided with 12 h light period (cool white fluorescent lamps, 25 uM $m^{-2}.s^{-1}$ 400–750 nm) and night temperature (15° C) coincided with the 12 h dark period. Percent germination was recorded every day for 15 days for all experiments. Un-germinated seeds from the salinity treatments were transferred to distilled water to study the recovery of germination, which was recorded each day for 7 days. The recovery percentage was determined by the following formula: (a–b)/(c–b) $\times$ 100; where a = the total number of seeds germinated after being transferred to distilled water, b = the total number of seeds germinated in saline solution and c = the total number of seeds. The rate of germination was estimated by using modified Timson's index of germination velocity: $\sum G/t$ – where $G$ is the percentage of seed germination at one day interval, and $t$ is the total germination period [2, 31]. The maximum number in our system could be 100. The higher value the more rapid the germination.

Data were analyzed using SPSS for window release 11.0. LSD test was used ($p < 0.05$) to determine significant differences among salinity treatments and species in seawater and NaCl separately.

## Results

Seed germination in both *S. salsa* and *A. centralasiatica* remained unaffected up to 20 $dS.m^{-1}$ NaCl and seawater treatments (Fig. 1). A further increase in salinity inhibited germination in both species; however, only 8 % of seeds germinated at highest salinity treatments in the case of *A. centralasiatica* in comparison to 56 % germination in *S. salsa* in NaCl solution. NaCl inhibited germination more in comparison to seawater solutions (Fig. 1).

In non-saline control, highest germination was recorded at 2 days in *S. salsa* and 6 days in *A. centralasiatica* (Fig. 2). Increase of salinity delayed the germination and at EC 50 $dS.m^{-1}$ salinity level first seed germinated after 9 days in NaCl solution and only 11 % seeds germinated in seawater solution in *A. centralasiatica.* While in *S. salsa* 46 % and 58 % seeds germinated in NaCl and seawater solutions, respectively.

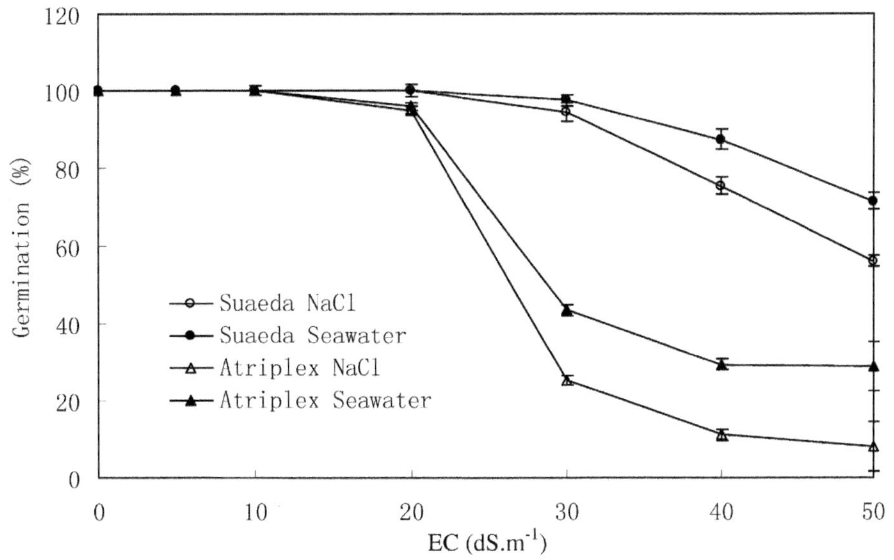


**Figure 1.** Mean final germination percentage of *Suaeda salsa* and *Atriplex centralasiatica* in various NaCl and seawater solutions. Bars represent means (± s.e., n=4).

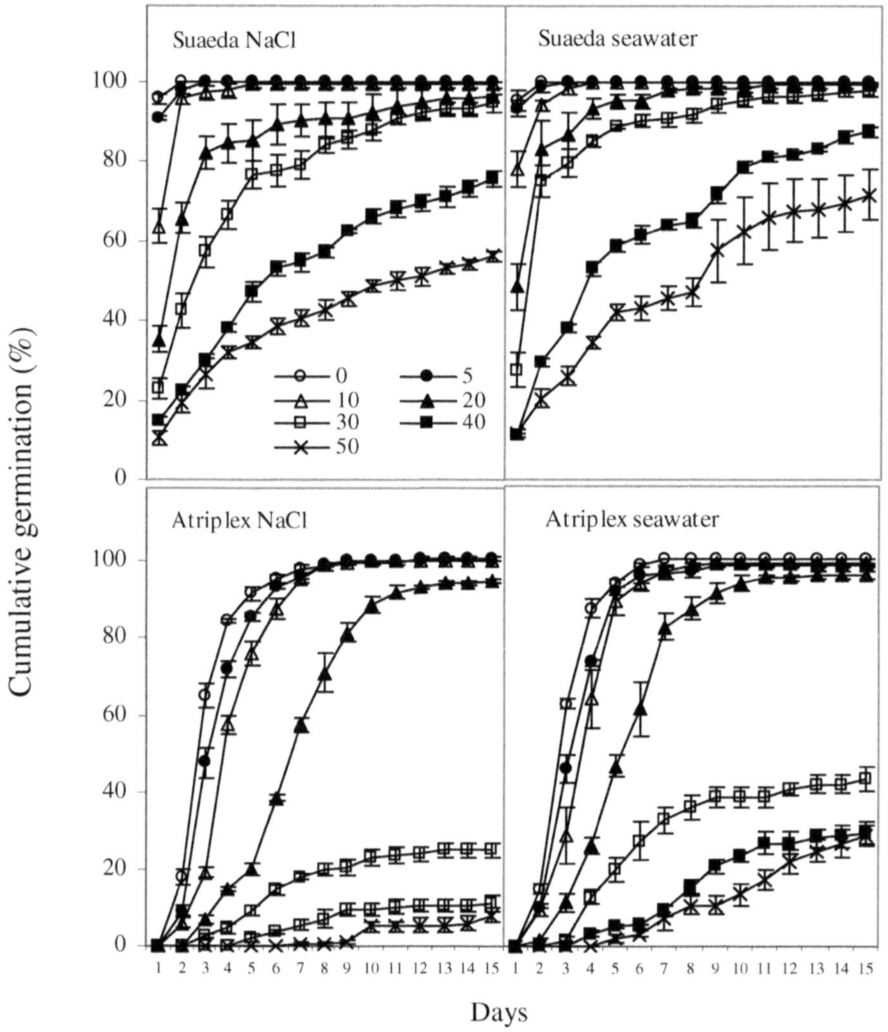


**Figure 2.** Cumulative mean germination percentage of *Suaeda salsa* and *Atriplex centralasiatica* over time in various NaCl and seawater concentrations. Bars indicate s.e. of means (n=4).

Rate of germination was highest in non-saline controls, and higher in *S. salsa* seeds than in *A. centralasiatica* seeds in all treatments and decreased with the increase of salinity (Fig. 3). Seeds of both species germinated slowly in NaCl solutions than in seawater solutions.

Un-germinated seeds from both NaCl and seawater solutions when transferred to distilled water recovered completely, but higher salinities affected the recovery of *A. centralasiatica*, and greater in seawater solutions than in NaCl solutions, for

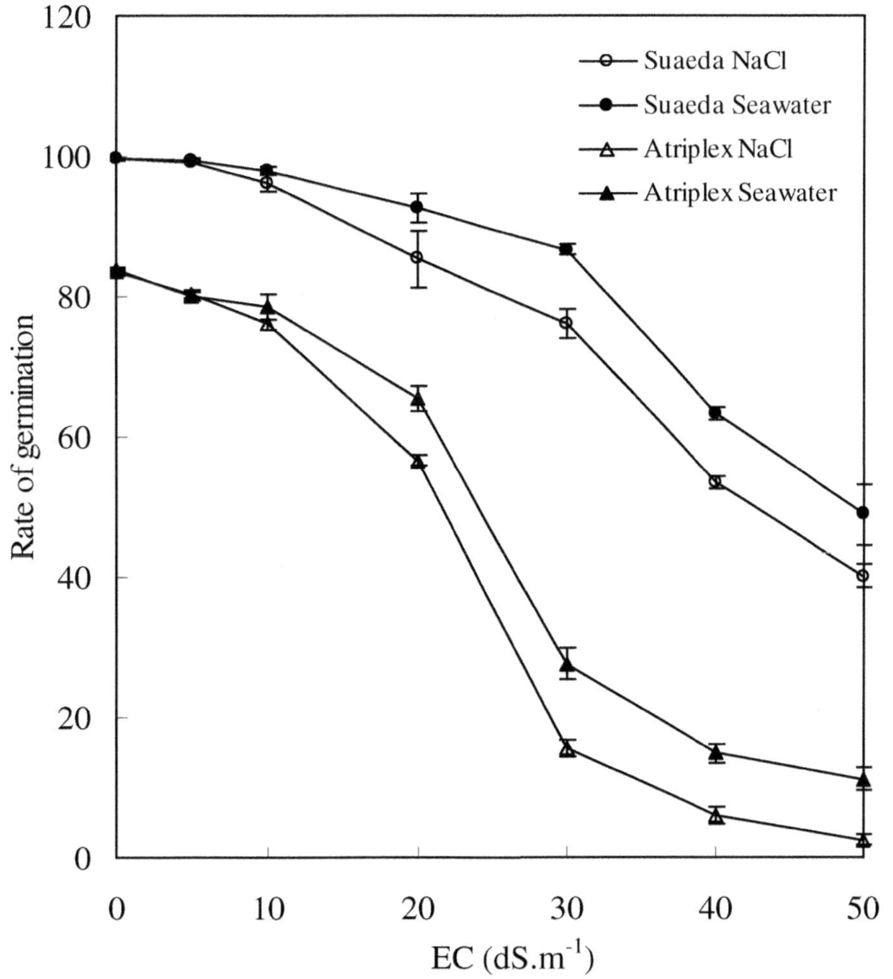


**Figure 3.** Rate of germination of *Suaeda salsa* and *Atriplex centralasiatica* seeds under various NaCl and seawater concentrations. Bars represent means (±s.e., n=4).

example, un-germinated seeds from 50 $dS.m^{-1}$, *S. salsa* can fully recovered at 1 day and 97 % in NaCl and 82 % in seawater of *A. centralasiatica* recovered at 7 days (Fig. 4).

## Discussion

The vegetation along the coast of Northern China is dominated by a stem succulent halophyte *Suaeda salsa* and a secreting halophyte *Atriplex centralasiatica*. Coastal communities are usually mono-specific and their distribution is controlled by inundation gradient and their frequency. *S. salsa* occupy low marsh habitat with high

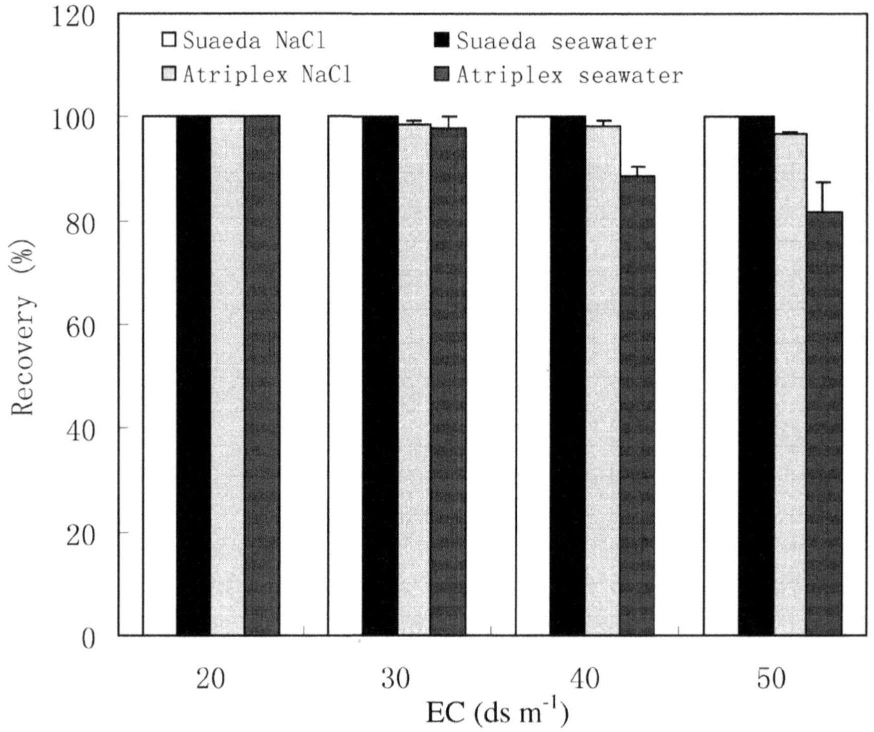


**Figure 4.** Mean recovery percent germination for *Suaeda salsa* and *Atriplex centrlasiatica* in distilled water under the treatments of various NaCl and seawater concentrations. Bars represent means (±s.e., n=4).

salinity and high moisture, while *A. centralasiatica* occupy a higher ground with infrequent inundation and low salinity and moisture conditions. This difference in ecological habitat has conferred varied ecological strategies to both species particularly at germination level. Seeds of *S. salsa* are highly salt tolerant and could germinate at seawater salinity, while few seeds of *A. centralasiatica* could germinate at that level. Halophytes vary in their ability to tolerate salinity at different stages of the life cycle. They are usually very highly salt tolerant while stored in the soil; however, their tolerance decreases at the germination and in most cases it increases again at the growth stage [4]. Seeds of leaf succulent species are highly tolerant to salinity [4]. Salinity plays an important role in determining the germination and survival of *Suaeda* spp [32] and it can tolerate high salinity during germination [32, 33]. The limit of salt tolerance of different species of *Suaeda* varies from 400–800 mM NaCl [22, 23, 34–36]. Secreting halophytes that could germinate above seawater salinity are few [7–11]. Most secreting halophytes show germination at NaCl concentrations ranging from 0.34–0.52 M NaCl [4, 37] while few of them have low salt tolerance during germination [12–14].

NaCl inhibited more seed germination in comparison to seawater solution both in *S. salsa* and *A. centralasiatica*. There is little information available on the effect of seawater on the germination of halophytes [7, 16, 17, 38, 39] and on the relative tolerance of seawater and NaCl solutions during seed germination [16, 17, 40]. Some results of combined salts effects on seed germination showed that the inhibitory effect of single salts can be considerably alleviated in natural soil systems by synergistic interactions between salts [18–21]. The increased seed germination in seawater may follow the same rule as combined salt effect because in seawater a lot of ions were included. However, seeds of *Limonium stocksii* [16] and *Salvadora persica* [17] were inhibited more by seawater; they also attributed this effect to seawater composition. This needs the further investigation.

Seeds of halophytes have the unique property of surviving extremely high salinity during the storage in the seed bank [41] and they germinate readily when soil salinity is reduced. However, recovery responses vary from one species to the other and against the level of salinity they are exposed to [27]. Seeds of *S. salsa* recovered completely when exposed to higher concentrations of NaCl and sea salt. Recovery of *A. centralasiatica* was also very high, except at 50 $dS.m^{-1}$ where 12–20% of seeds failed to recover. A substantial recovery from germination occurred at the NaCl concentrations up to 600 mM NaCl in *Halogeton glomeratus* [42], *Sarcobatus vermiculatus* [43], *Suaeda moquinii* [36] and *Triglochin maritima* [44].

*Suaeda salsa* and *A. centralasiatica* are highly salt tolerant halophyte species where *S. salsa* is more salt tolerant than *A. centralasiatica*. Seeds of both species could germinate at seawater level but more germination was recorded in the seeds of *S. salsa*. Seed germination of *S. salsa* was more rapid in comparison to *A. centralasiatica*. Both species have a high ability to survive under extreme conditions; the un-germinated seeds can recover completely when the salinity stress was removed. Seed germination was inhibited more by NaCl than seawater. The literature suggests that the germination of halophyte seeds have differential response to seawater and NaCl. Some reports indicate that seawater inhibits more seed germination, while others believe that it is NaCl. Further studies will be carried out to understand the difference between NaCl and seawater effects on germination.

## References

1 Ungar IA (1991) *Ecophysiology of Vascular Halophytes*. CRC Press, Boca Raton, Florida
2 Khan MA, Ungar IA (1984) The effect of salinity and temperature on germination of polymorphic seeds and growth of *Atriplex triangularis* Willd. *Am J Botany* 71: 481–489
3 Ungar IA (1995) Seed germination and seed bank ecology in halophytes. In: J Kigel, G Galili (eds): *Seed Development and Seed Germination*. Marcel Dekker, New York 599–628
4 Khan MA, Gul B (2002) Some ecophysiological aspects of seed germination in halophytes. In: X Liu, M Liu (eds): *Halophytes Utilization and Regional Sustainable Development of Agriculture*. Meteorological Press. Beijing 56–68
5 Chapman VJ (1960) *Salt marshes and salt deserts of the world*. Interscience Publishers, New York
6 Khan MA, Gul B (1998) High salt tolerance in germinating dimorphic seeds of *Arthrocnemum indicum*. *Int J Plant Sci* 159: 826–832

7 Woodell SRJ (1985) Salinity and seed germination patterns in coastal plants. *Vegetation* 61: 223–229

8 Khan MA (1991) Studies on germination of *Cressa cretica* L. seeds. *Pakistan J Weed Sci Res* 4: 89–98

9 Ungar IA (1967) Influence of salinity and temperature on seed germination. *Ohio J Sci* 67: 120–123

10 Binet P (1965) Action de divers rythmes thermiques journaliers sur la germination de semences de *Triglochin maritimum* L. *Bulletin Societe. Normandie (Caen), 10e series.* 6: 99–102

11 Ignaciuk R, Lee JA (1980) The germination of four annual strand-line species. *New Phytologist* 84: 581–587

12 Mahmoud A, El Sheikh AM, Abdul Baset S (1983) Germination of two halophytes: *Halopaplis perfoliata* and *Limonium axilare* from Saudi Arabia. *J Arid Env* 6: 87–98

13 Ladiges PY, Foord PC, Willis RJ (1981) Salinity and water logging tolerance of some populations of *Melaleuca ericifolia* Smith. *Australian J Ecol* 6: 203–215

14 Fernandez G, Johnston M, Olivares PA (1985) Rol del pericarpio de *Atriplex repanda* en la germinacion. III. Estudio histological y quimico del pericarpio. *Phyton* 45: 165–169

15 Ungar IA (1978) Halophyte seed germination. *Botanical Review* 44: 233–264

16 Zia S, Khan MA (2002) Comparative effect of NaCl and seawater on seed germination of *Limonium stocksii. Pakistan J Botany* 34: 345–350

17 Joshi AJ, Mali BS, Hingalajia H (1995) Halophytes – A good source of forage production under salt stress conditions l. *Salvadora persica.* In: MA Khan and IA Unger (eds): *Biology of Salt Tolerant Plants.* Department of Botany, University of Karachi 353–360

18 Duan D, Liu X, Feng F, Li C (2003) Effects of salinities on seed germination of halophytes *Suaeda salsa. Chinese Agricultural Science Bulletin* 19(6): 168–172

19 Duan D, Liu X, Khan MA, Gul B (2004) Effects of salt and water stress on the germination of *Chenopodium glaucum* L. seed. *Pakistan J Botany* 36: 793–800

20 Al-Jibury LK, Clor MA, Talabany D (1986) Effects of certain salts and their combinations on germination and seedling development of *Securigera securidaca* L. *Arabic Gulf J Sci Res* 4: 5–16

21 Zhang J, Li J, Xing S, Song Y, Chi J (2003) Experiment of seed germination of *Rhus chinensis* under salt stress. *J Northeast Forestry Uni* 31: 79–80

22 Ungar IA (1962) Influence of salinity on seed germination in succulent halophytes. *Ecology* 3: 329–335

23 Ungar IA, Capilupo F (1969) An ecological life history study of *Suaeda depressa* (Pursh.) Wats. *Advancing Frontiers of Plant Sciences* 23: 137–158

24 Clarke LD, Hannon NJ (1970) The mangrove swamp and saltmarsh communities of the Sydney District. III: plant growth in relation to salinity and water logging. *J Ecol* 58: 351–369

25 Boorman LA (1967) Experimental studies in the genus *Limonium.* Ph.D. dissertation, University of Oxford, UK

26 Boorman LA (1968) Some aspects of the reproductive biology of *Limonium vulgare* Mill. and *L. humile*_mill. *Annals of Botany* 32: 803–824

27 Keiffer CW, Ungar IA (1995). Germination responses of halophyte seeds exposed to prolonged hypersaline conditions. In: MA Khan and IA Unger (eds): *Biology of Salt Tolerant Plants.* Department of Botany, University of Karachi 43–50

28 Zhao K, Li F (eds) (1999) *Halophytes in China.* Science Press. Beijing

29 Zhao K, Fan H, Ungar IA (2002) Survey of halophyte species in China. *Plant Sci* 163: 491–498

30 Bai Yajun, Liu Xiaojing, Li Weiqiang (2003) Primary analysis of four salt tolerant plants growing in Hai-He Plain, China. In: H Lieth, M Mochtchenko (eds): *Tasks for vegetation science-38, Cash crop halophytes recent studies*. Kluwer Academic Publishers, Dordrecht, the Netherlands 135–138
31 Timson J (1965) New methods of recording germination data. *Nature* 207: 216–217
32 Sheikh KH, Mahmood K (1986) Some studies on field distribution and seed germination of *Suaeda fruticosa* and *Sporobolus arabicus* with reference to salinity and sodicity of the medium. *Plant and Soil* 94: 333–340
33 Rajpurohit KS, Sen DN (1977) Soil salinity and seed germination under water stress. *Transaction of Indian Society of Desert Technology and University Center of Desert Studies* 2: 106–110
34 Okusanya OT (1979) An experimental investigation into the ecology of some maritime cliff species. II. Germination studies. *J Ecol* 67: 293–304
35 Yokoishi T, Tanimoto S (1994) Seed germination of the halophyte *Suaeda japonica* under salt stress. *J Plant Res* 107: 385–388
36 Khan MA, Gul B, Weber DJ (2001) Germination of dimorphic seeds of *Suaeda moquinii* under high salinity stress. *Australian J Botany* 49: 185–192
37 Khan MA, Rizvi Y (1994) The effect of salinity, temperature and growth regulators on the germination and early seedling growth of *Atriplex griffithii* Moq. var. *stocksii* Boiss. *Canadian J Botany* 72: 475–479
38 Rivers WG, Weber DJ (1971) The influence of salinity and temperature on seed germination in *Salicornia bigelovii*. *Physiologia Plantrum* 24: 73–75
39 McMillan C (1988) Seed germination and seedling development of *Halophphila decipiens* Ostenfeld from Panama. *Aquatic Botany* 31: 169–176
40 Tirmizi SAS, Khan KM, Qadir SA (1993) Study on salt tolerance of *Hippophae rhamnoides* L., during germination. *Pakistan J Sci Ind Res* 36: 252–257
41 Khan, MA, Ungar IA (1997) Effect of thermoperiod on recovery of seed germination of halophytes from saline conditions. *American J Botany* 84: 279–283
42 Khan MA, Gul B, Weber DJ (2001) Seed germination characteristics of *Halogeton glomeratus*. *Canadian J Botany* 79: 1189–1194
43 Khan MA, Gul B, Weber DJ (2002) Effect of temperature, and salinity on the germination of *Sarcobatus vermicullatus*. *Biologia Plantarum* 45: 133–135
44 Khan MA, Ungar IA (1999) Seed germination and recovery of *Triglochin maritima* from salt stress under different thermoperiods. *Great Basin Naturalist* 59: 144–150

Biosaline Agriculture and Salinity Tolerance in Plants
Edited by M. Öztürk, Y. Waisel, M.A. Khan and G. Görk


# Salt effect on growth, photosynthesis, seed yield and oil composition of the potential crop halophyte *Cakile maritima*

Ahmed Debez[1], Wael Taamalli[2], Dhouha Saadaoui[1], Zeineb Ouerghi[1], Mokhtar Zarrouk[2], Bernhard Huchzermeyer[3] and Chedly Abdelly[1]

[1] *Laboratoire d'Adaptation des Plantes aux Stress Abiotiques, INRST, BP 95, Hammam-Lif, 2050, Tunisia*
[2] *Laboratoire de Caractérisation et Qualité de l'Huile d'Olive, INRST, BP 95, Hammam-Lif, 2050, Tunisia*
[3] *Institut für Botanik, Universität Hannover, Herrenhäuserstr. 2, 30419 Hannover, Germany*

## Introduction

Salinity is an extending environmental issue which compromises the long-term sustainability of agriculture, especially in the coastal semi-arid areas [1]. This is the case in Tunisia, where the semi-arid Mediterranean climate prevails (mean annual precipitation of 200–700 mm). Subsequently, around 10 % of the whole territory would be salt-affected [2]. Halophytes have evolved a wide range of attributes (morphological, physiological and biochemical) allowing them to tolerate the presence of salt in the medium [3]. Besides, several studies suggest that these plants are potentially useful for ecological and economical purposes [4].

Intracellular salt flux control is one of the major salt tolerance determinants, involving salt exclusion and/or compartmentation. P and V $H^+$-ATPases (respectively localized at the plasma membrane and tonoplast) provide energy for $Na^+/H^+$ antiporters, thus allowing sodium active transport away from the cytoplasm [5]. The bi-directional transport of sodium insures ion homeostasis, cell turgor, as well as the metabolic functioning [6]. On the other hand, impairment of the photosynthetic activity greatly accounts for growth restriction of non-halophytes under salinity [7]. Depressive effects of salinity are thought to arise from stomatal and/or non stomatal limitations (i.e., stomatal closure and/or damage to Calvin cycle enzymes) [8].

*Cakile maritima* (*Brassicaceae*) is an annual fleshy halophyte which colonizes the sandy beaches of the Tunisian littoral. This study aims to characterize the plant response to long-term salt treatments (0–500 mM NaCl), using physiological (growth, water status, mineral nutrition) and biochemical ($H^+$-ATPase activity and photosynthetic capacity) criteria. Changes in seed yield and seed oil characteristics under salinity are also assessed.

## Material and methods

Mature seeds of *C. maritima* were harvested on sandy beaches of Raoued (20 km to the north of Tunis). Seedlings were grown in pots filled with inert sand in a glasshouse (16 h/8 h light/dark regime; 60 % relative humidity; 300 $\mu$mol·m$^{-2}$·s$^{-1}$ photosynthetic active radiation – PAR; $22 \pm 1°$ C temperature). Irrigation was performed with a Long Ashton nutrient solution [9]. Four week-old plants were progressively submitted to increasing salinities (0–500 mM NaCl) for 6 weeks. The same experience was repeated until the complete maturation of seeds. At the final harvest, physiological parameters were determined (i.e., dry weight, leaf number and area, leaf succulence ratio, leaf ion status). Yield components assessed were seed yield, seed mass, seed viability, and seed oil content.

Expanded leaves situated on the fifth node from the shoot top were used for photosynthetic and $H^+$-ATPase activity measurements. Leaf gas exchanges were measured with a portable photosynthesis system (LCi, ADC Bioscientific Ltd., UK) at 2,500 $\mu$mol.m$^{-2}$.s$^{-1}$ PAR (saturating light). Ribulose-biphosphate carboxylase/oxygenase (Rubisco; EC 4.1.1.39) activity was spectrophotometrically assayed ($\lambda$ = 340 nm) [10]. Vacuolar and plasma membrane (respectively, V and P) $H^+$-ATPase activities were assayed on isolated chloroplasts, using [$\gamma-^{32}$P] ATP (1 MBq) (Hartmann Analytik, Braunschweig, Germany) [11]. Seed total lipids were extracted [12] and triacylglycerols (TAG) were separated by thin layer chromatography (TLC), using silica gel plates (Merck G 60) [13]. Fatty acid methyl esters were quantified by adding heptadecanoic acid (17:0) as an internal standard. Results are the means of three samples. A one way ANOVA was achieved to compare the mean values, using the SPSS statistical program ($P < 0.05$). In case of significant differences, Duncan *post hoc* tests were performed.

## Results and discussion

Moderate salinities (50–100 mM NaCl) were optimal for the plant growth, since improving whole plant dry weight (+24 % at 100 mM NaCl) (Fig. 1A). No significant growth decrease occurred in the 200–300 mM NaCl range (ca. 90 % of control values), and the plant was able to survive, even at a salinity close to that of seawater (500 mM NaCl). These data corroborate previous investigations on other halophytes, showing sub-optimal growth in mediums lacking salt [14, 15]]. Leaves largely accounted for the plant response pattern, since their dry weight and number were significantly stimulated at optimal salinities (50–100 mM NaCl) (Figs 1A and 1B). Leaf water status, evaluated by leaf succulence ratio, was significantly enhanced by salt treatments (Fig. 2A), and remained higher than the control values, even at 500 mM NaCl. The improvement of leaf hydration under salinity was concomitant with the accumulation of high amounts of $Na^+$, and at a lesser content of $Cl^-$ (Fig. 2B) (1.8 and 3.8 mmol.g$^{-1}$ DW, respectively). Since salt treatment did not impair leaf hydration, most of $Na^+$ ions transported in leaves might have been removed from the leaf apoplast and efficiently compartmentalized by cells for water retention. Salinity restricted the plant nutrient uptake, leading to a significant decrease in leaf $K^+$

contents (Fig. 2B). The same tendency was observed for $Ca^{2+}$ and $Mg^{2+}$ (data not shown). The salt-induced reduction of growth could be a consequence of nutritional imbalance. Moreover, despite $Na^+$ is a cheap osmoticum for halophytes, an excess of this ion over $K^+$ can inhibit several metabolic processes.

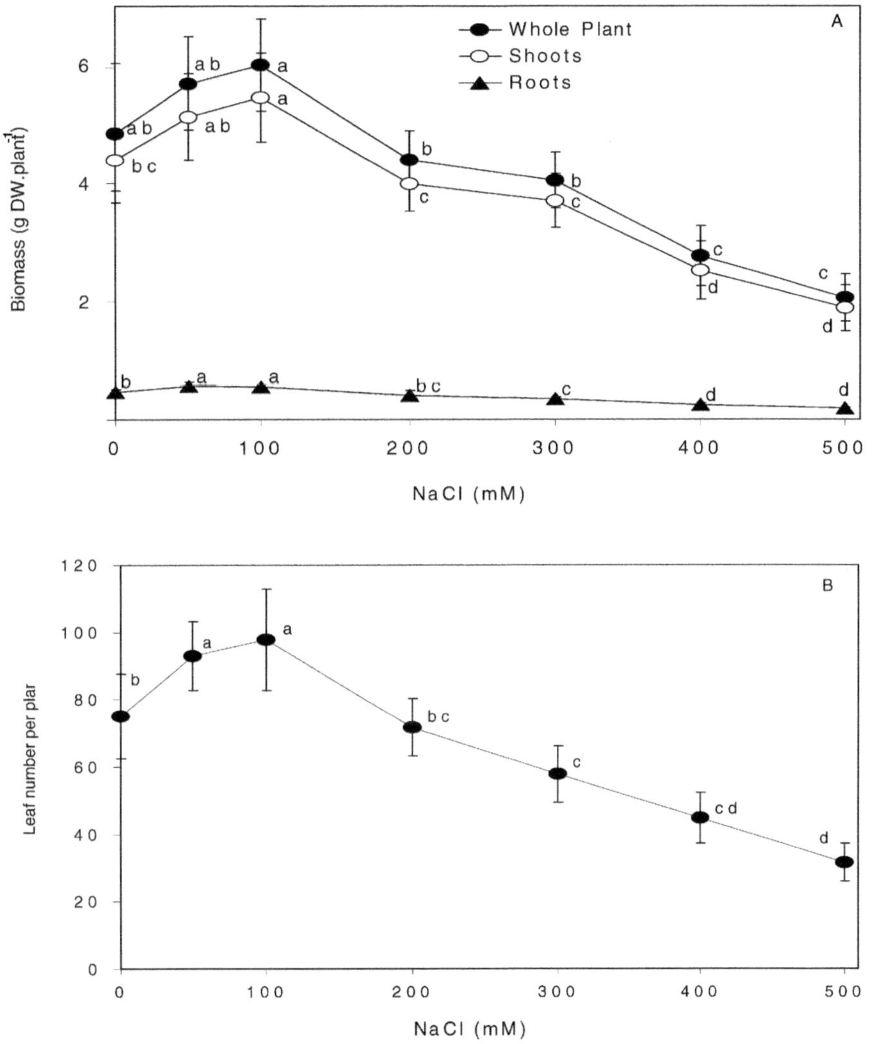


**Figure 1.** Effect of NaCl on growth of *C. maritima*. (**A**) Biomass production of the whole plant and the different organs. (**B**) Leaf number per plant. Means of 18 replicates ± SE. (Values within each salt treatment marked with at least one same letter are not significantly different at $P < 0.05$.)

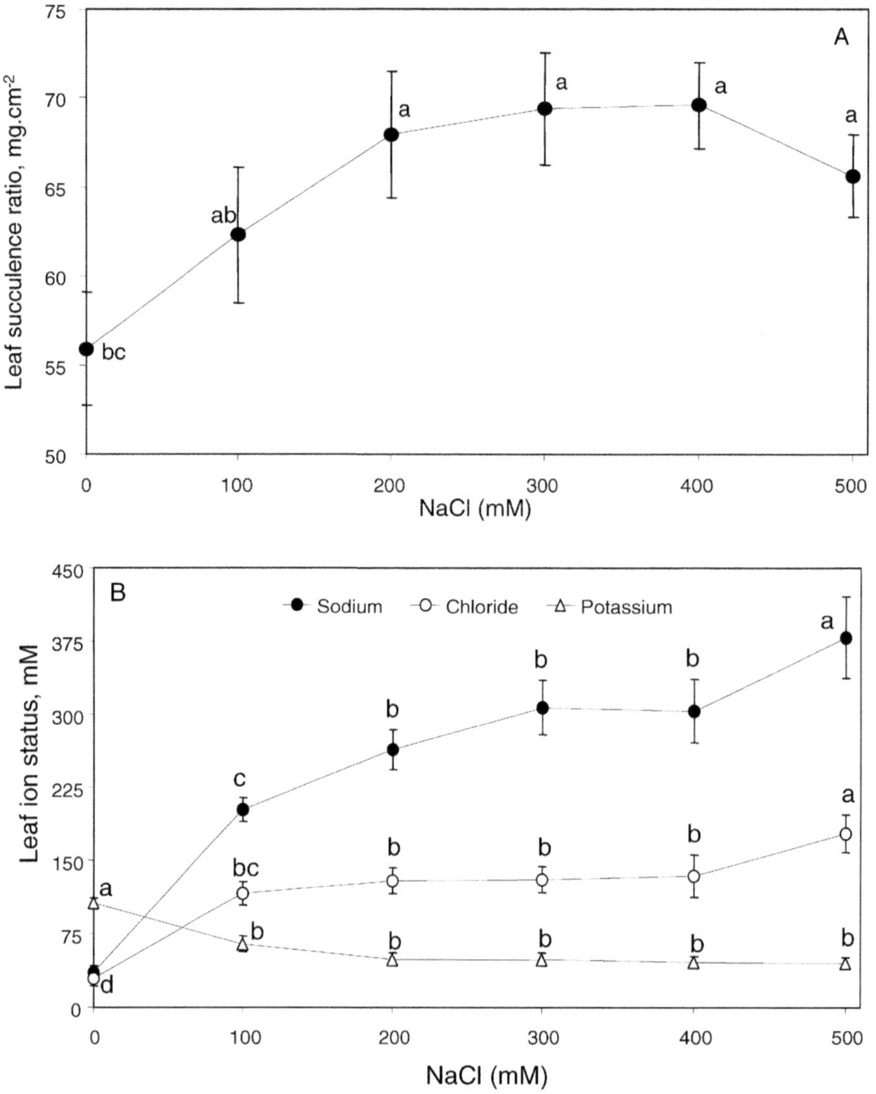


**Figure 2.** Effect of NaCl on leaf water status and mineral nutrition of *C. maritima*. (**A**) Leaf succulence ratio. (**B**) Leaf ion contents. Means of 18 replicates ± SE. (Values within each salt treatment marked with at least one same letter are not significantly different at $P < 0.05$.)

Combining the results relative to the leaf water status and salt accumulation of salt-treated *C. maritima* provide indirect evidence for the existence of salt compartmentation mechanisms within leaf cells (i.e., inclusive strategy). This assumption was confirmed by (*i*) the strong stimulation of V $H^+$-ATPase activity up to 300 mM NaCl (+80 %/control) (Fig. 3A) and (*ii*) the absence of anatomical structures responsible for salt exclusion at the leaf surface. NaCl concentrations in the 300–500 mM range significantly promoted P $H^+$-ATPase activity, suggesting that the exclusive pattern may take place at high salinities (Fig. 3B). Keeping sodium and chloride away from cytosol (using inclusive strategy) is of vital importance for dicotyledonous halophytes lacking morphological structures of salt excretion at their leaf surface [16]. Owing to their catalyzer role, proton pumps enable both vacuolar and plasma membrane antiporter functioning, and play therefore, a major role in salt tolerance [17]. In addition, the overexpression of $Na^+/H^+$ antiporters plants has been reported to improve the performance of several species in saline conditions [18].

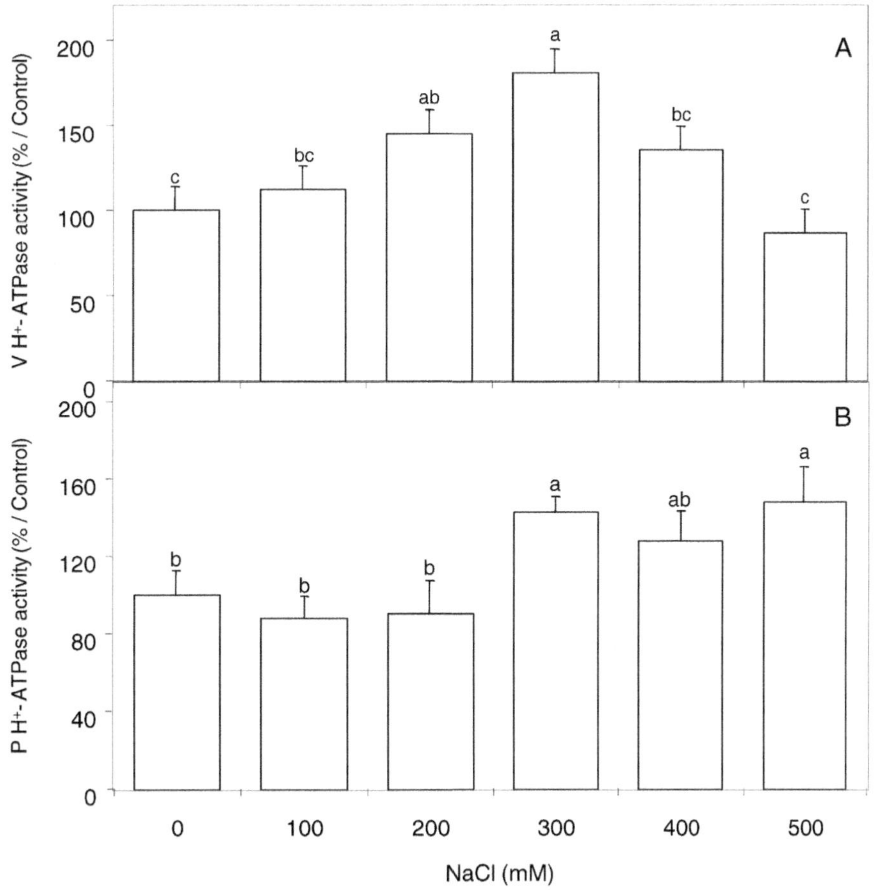


**Figure 3.** Effect of NaCl on $H^+$-ATPase activity (%/Control) of *C. maritima*. (**A**) Changes in vacuolar V $H^+$-ATPase activity. (**B**) Changes in plasma membrane P $H^+$-ATPAse activity. Means of three replicates ± SE. (Values within each salt treatment marked with at least one same letter are not significantly different at $P < 0.05$.)

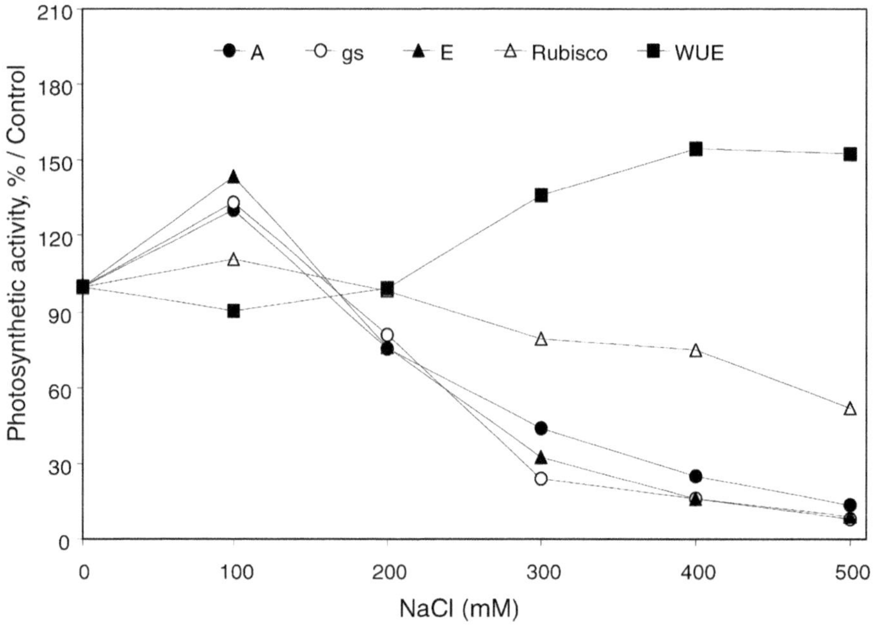


**Figure 4.** Effect of NaCl on photosynthetic activity of *C. maritima*. Results of gas exchanges are the means of 10 replicates. Results of Rubisco activity are the means of three replicates.

Both stomatal and non stomatal components of photosynthesis were improved at optimal salinity for growth. $CO_2$ assimilation rate (A), stomatal conductance ($g_s$), and transpiration rate (E) were 30–40 % higher at 100 mM NaCl, while specific activity of Rubisco was augmented by ca. 10 % (Fig. 4). Supra-optimal salinities impaired photosynthetic activity, but this depressive effect was more pronounced on stomatal conductance than on enzyme activity (respectively 15 % and 75 % of the control values at 400 mM NaCl). Former studies showed that stomatal limitation accounted for the reduction of photosynthesis in salt-treated plants [8]. In *C. maritima*, stomata closure was associated with reduced transpiration rate (E), leading to higher water-use efficiency (+ ca. 50 % at 500 mM NaCl). No salt-induced shift in the photosynthetic pathway ($C_3$ to $C_4$) was observed, since phosphoenolpyruvate carboxylase (PEPC, EC 4.1.1.39) activity remained lower than Rubisco one, irrespective of salt treatment (data not shown).

Optimal salinities for growth and photosynthesis promoted significantly seed production (+50 % in the 50–100 mM NaCl range) (Fig. 5A). This parameter was more affected than plant growth at high salinities (respectively 21 % and 84 % of the control values at 300 mM NaCl), likely resulting from a reduction of flower production and/or a decrease of their fertility [19]. The mean mass of individual seed decreased significantly in the presence of salt in *C. maritima* (Fig. 5A), indicating that assimilate allocation to seeds was more restricted by salt than seed initiation. Seeds harvested from plants exposed to mild salinities (50–200 mM NaCl) displayed

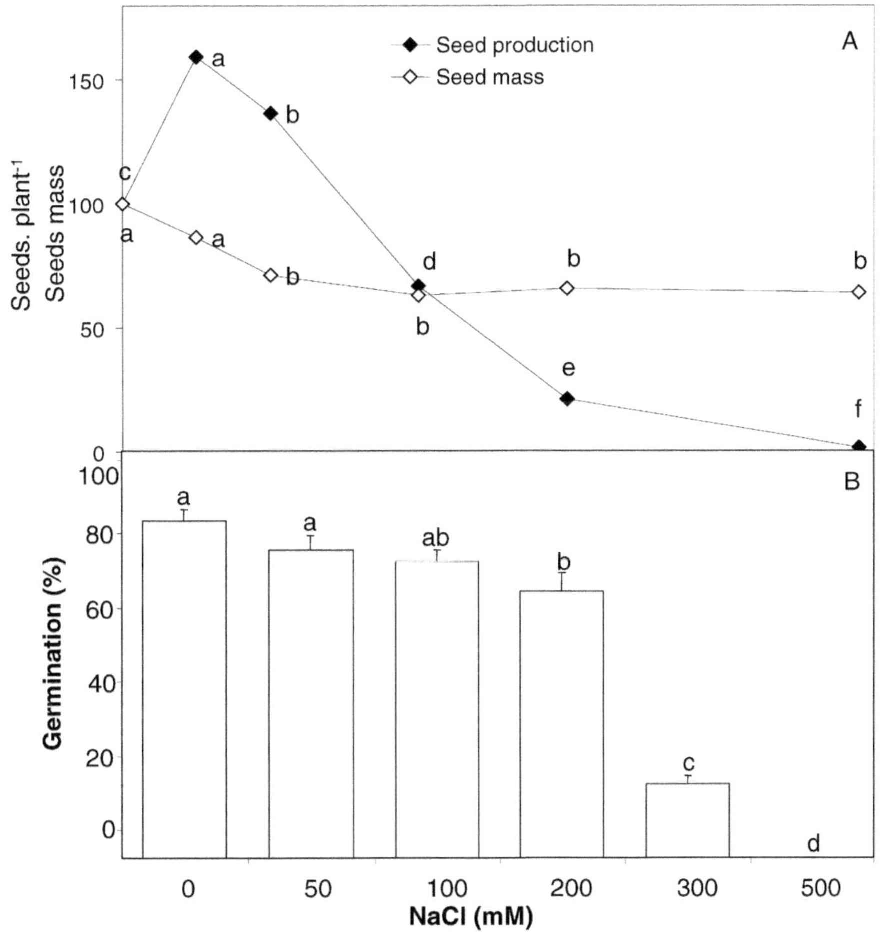


**Figure 5.** Effect of NaCl on the reproductive capacity of *C. maritima*. (**A**) Seed production per plant (means of 12 replicates) and individual seed mass (mg) (means of 300 replicates), expressed as % of the control. (**B**) Germination capacity (%) in distilled water of seeds harvested from plants exposed to increasing salinities. Means of three replicates ± SE. (Values within each salt treatment marked with at least one same letter are not significantly different at $P < 0.05$.)

high germination rates (up to 80 %), contrasting with those produced under high salt levels (Fig. 5B). Increasing salinities led to both quantitative and qualitative changes in the seed oil characteristics. Seed oil content (on a dry weight basis) was positively correlated with the medium salinity (respectively 30 % and 28 % at 100 mM and 500 mM NaCl). Seed oil content seemed also to be unaffected by salinity in the oleaginous halophyte *Lesquerella fendleri* [20], while decreasing in sunflower [21]. Erucic acid (22:1) level increased markedly, reaching 26 % at 500 mM NaCl (two-fold higher than control values). This trend was concomitant with a significant decrease in oleic acid (18:1) level (ca. 45 % of the control value at 500 mM NaCl). In our conditions, higher erucic acid in salt treated *C. maritima* was associated with increased 22:1/18:1 ratio (0.39 and 1.36, respectively for the control and 500 mM NaCl plants), likely mediating elongases, which are known to catalyze the formation of long fatty acids (such as erucic acid), using oleic acid as initial substrate [22].

In summary, the present study shows that moderate salinities are required by *C. maritima* to express maximal growth and seed production potentialities, in relation with the concomitant involvement of several processes at different levels. Further field experiments are necessary to confirm the economic potential of this promising halophyte.

*Acknowledgements*
This work was supported by a sandwich scholarship granted by the German Academic Exchange Service (Deutscher Akademischer Austausch Dienst, DAAD). We wish to thank Professor Ahlert Schmidt (Institut für Botanik, University of Hannover, Germany) for his kind collaboration and for providing technical facilities.

## References

1 Horie T, Schroeder JI (2004) Sodium transporters in Plants. Diverse genes and physiological functions. *Plant Physiol* 136: 2457–2462
2 Hachicha M, Job JO, Mtimet A (1994) Les sols salés et la salinisation en Tunisie. *Sols de Tunisie* 5: 271–341
3 Glenn E, Brown JJ, Blumwald E (1999) Salt-tolerant mechanism and crop potential of halophytes. *Crit Rev Plant Sci* 18: 227–255
4 Glenn EP, O'Leary JW, Watson MC, Thompson TL, Kuehl RO (1991) *Salicornia bigelovii* Torr.: an oilseed halophyte for seawater irrigation. *Science* 251: 1065–1067
5 Zhu JK (2001) Plant salt tolerance. *Trends Plant Sci* 6: 66–71
6 Aharon GS, Apse MP, Duan S, Hua X, Blumwald E (2003) Characterization of a family of vacuolar $Na^+/H^+$ antiporters in *Arabidopsis thaliana. Plant Soil* 253: 245–256
7 Munns R (2002) Comparative physiology of salt and water stress. *Plant Cell Environ* 25: 239–250
8 Meloni DA, Oliva MA, Martinez CA, Cambraia J (2003) Photosynthesis and activity of superoxide dismutase, peroxidase and glutathione reductase in cotton under salt stress. *Environ Exp Bot* 49: 69–76
9 Abdelly C, Lachaâl M, Grignon C, Soltani A, Hajji M (1995) Association épisodique d'halophytes strictes et de glycophytes dans un écosystème hydromorphe salé en zone semiaride. *Agronomie* 15: 557–568
10 Sato FK, Nishida K, Yamada Y (1980) Activities of carboxylation enzymes and products of $^{14}CO_2$ fixation in photoautotrophically cultured cells. *Plant Sci Lett* 20: 91–97
11 Koyro H-W, Stelzer R, Huchzermeyer B (1993) ATPase activities and membrane fine structure of rhizodermal cells from *Sorghum* and *Spartina* roots grown under mild salt stress. *Bot Acta* 106: 110–119
12 Allen C, Good P (1971) Acyl lipids in photosynthetic systems. *Methods in Enz* 23: 523–547
13 Mangold HK (1964) Thin layer chromatography of lipids. *JAOCS* 47: 726–773
14 Bajji M, Kinet JM, Lutts S (1998) Salt stress effects on roots and leaves of *Atriplex halimus* L. and their corresponding callus cultures. *Plant Sci* 137: 131–142

15 Short DC, Colmer TD (1999) Salt tolerance in the halophyte *Haloscaria pergranulata* subsp. *pergranulata*. *Ann Bot* 83: 207–213

16 Véry A-A, Robinson MF, Mansfield TA, Sanders D (1998) Guard cell cation channels are involved in $Na^{+}$-induced stomatal closure in a halophyte. *Plant J* 14: 509–521

17 Zhu J-K (2003) Regulation of ion homeostasis under salt stress. *Curr Opin Plant Biol* 6: 441–445

18 Venema K, Belver A, Marin-Manzano MC, Rodriguez-Rosales MP, Donaire JP (2003) A novel intracellular $K^{+}/H^{+}$ antiporter related to $Na^{+}/H^{+}$ antiporters is important for $K^{+}$ ion homeostasis in plants. *J Biol Chem* 278: 22453–22459

19 Khatun S, Flowers TJ (1995) The estimation of pollen viability in rice. *J Exp Bot* 46: 151–154

20 Dierig DA, Grieve CM, Shannon MC (2003) Selection for salt tolerance in *Lesquerella fendleri* (Gray) S. Wats. *Indust Crops Prod* 17: 15–22

21 Flagella Z, Giuliani MM, Rotunno T, Di Caterina R, De Caro A (2004) Effect of saline water on oil yield and quality of a high oleic sunflower (*Helianthus annuus* L.) hybrid. *Eur J Agron* 21: 267–272

22 Katavic V, Mietkiewska E, Barton DL, Giblin EM, Reed DW, Taylor DC (2002) Restoring enzyme activity in nonfunctional low erucic acid *Brassica napus* fatty acid elongase 1 by a single amino acid substitution. *Eur J Biochem* 269: 5625–5631

Biosaline Agriculture and Salinity Tolerance in Plants
Edited by M. Öztürk, Y. Waisel, M.A. Khan and G. Görk


# Effect of nitrogen deficiency, salinity and drought on proline metabolism in *Sesuvium portulacastrum*

Dorsaf Messedi[1], Ines Slama[1], Nahla Laabidi[1], Tahar Ghnaya[1], Arnould Savoure[2], Abdelaziz Soltani[1] and Chedly Abdelly[1]

[1] *Laboratoire d'Adaptation des Plantes aux Stress Abiotiques, INRST, BP 95, Hammam-Lif, 2050, Tunisia*
[2] *Unité Physiologie Cellulaire et Moléculaire des Plantes, CNRS-UMR 7632, 4 Place Jussieu, Paris, France*

## Introduction

Drought and high salinity are responsible for large decreases in crop productivity all over the world [1]. These losses of crop yield result from limitations of plant development through excessive ion accumulation, water deficit and mineral deficiencies [2]. Under these prevalent stresses, tolerant plants adopt various strategies with a wide range of biochemical to physiological and morphological adaptations [3]. Morphological ones include modifications in growth and allocation of assimilates towards roots for an efficient exploitation of soil nutrients [4]. The physiological strategy is represented by a higher selectivity for $K^+$ over $Na^+$ [5], an increase in $K^+$-use efficiency [6], and the synthesis of organic osmolytes, with low molecular weight, for osmo-protection [7]. These osmolytes are sugars, polyols, amino acids, tertiary and quarternary ammonium, and tertiary sulphonium compounds [8].

The accumulation of compatible solutes induces a decrease in the water potential and allows additional water to be taken up from the environment [9]. In our study we focused on proline accumulation. Proline is commonly referred to as compatible solute in many eubacteria, algae, and higher plants [10]. The accumulation of proline is due primarily to *de novo* synthesis [11, 12], secondary to a reduced rate of catabolism [11], and finally to specific transport systems that distribute proline to the locations of need [13]. Two possible pathways of proline synthesis have been shown in plants. One is using glutamate (Glu) and the second is using ornithine (Orn) as a precursor [10]. Proline degradation in plants takes place in mitochondria and is catalysed by proline dehydrogenase (ProDH), also named proline oxidase [14]. Proline degradation has been shown to be inhibited under water and salt stresses. Both a decrease in ProDH mRNA level [15] and ProDH activity [16] result in proline accumulation.

In the present study, we investigated *S. portulacastrum* response to the availability of nitrogen in the presence of NaCl or under water stress. We measured δ-OAT and

ProDH activities to evaluate the relationship between enzyme activities and proline concentrations in leaves.

## Material and methods

### *Culture*

3 cm long stem segments with one node and two opposite leaves were taken from mother plants of *Sesuvium portulacastrum*, disinfected for 5 min in saturated calcium hypochlorite solution, and rinsed abundantly with distilled water. They were then placed for 7 days in an aerated Hewitt [17] solution diluted 10 times, supplemented with Fe K EDTA [18] and micronutrients [19]. Rhizogenesis took place during this week.

To determine the response of *S. portulacastrum* to the availability of nitrogen in presence of 400 mM NaCl (increased by 100 mM daily to reach the maximum salinity levels [20]), plants were submitted to a discontinuous nitrogen supply. After 35 days of pretreatment, an initial harvest was achieved. Plants were divided in two lots: in the first one, plants were cultivated on a complete nitrogen mode, 14.4 mM (+N), in the second one plants were subjected to limiting nitrogen supply 0.28 mM (–N). After 5 weeks of treatment, one lot of deficient plants was transferred on a complete nitrogen mode ($\pm$ N plants). Shoots and roots were weekly harvested during 70 days.

The second experiment aimed at determining the response of this halophyte to water deficit. Plants cultivated individually in pot filled with limono-sandy soil, were divided in two lots: the first one was irrigated with tap water at 100 % field capacity (FC) corresponding to control plants, and the second one at only 25 % FC (dehydrated plants). After 16 days of treatment, one lot of dehydrated plants was rewatered at 100 % FC. All cultures were carried out in a greenhouse with a 14 h photoperiod. Mean temperature and relative humidity were respectively $30 \pm 5°$ C, $55 \pm 5$ % day and $16 \pm 2°$ C, $90 \pm 5$ % night. Shoots and roots were harvested every 4 days during 40 days.

### *Plant analysis and enzymatic assay*

During the harvests, shoot and root dry weights were measured, after desiccation for 48 h at 60° C. Reduced nitrogen was measured according to the Kjeldahl method. Proline was extracted and estimated by the method suggested by Bates et al. [21]. Frozen leaves (three samples of approximately 1 g FW per treatment) were grounded to a fine powder in a chilled mortar and pestle in the presence of PVP (0.2 g/g FW), and then homogenized in an appropriate extraction buffer. Ratios for buffer volume/g FW were 2:1. All operations were carried out at 4° C. Extraction buffer of δ-OAT (EC 2.6.1.13) consisted of 100 mM K-Pi buffer (pH 7.9), 1 mM EDTA, 15 % glycerol and 10 mM 2-mercaptoethanol. The extract was centrifuged at $15,000 \times g$ for 15 min. Extraction buffer used for ProDH (EC 1.5.99.8) was 50 mM Tris-HCl buffer (pH 7.4) containing 7 mM $MgCl_2$, 0.6 M KCl, 3 mM EDTA and 1 mM DTT. The extract was centrifuged at $39,000 \times g$ for 20 min [22]. δ-OAT activity was assayed with ninhydrin according to Kim et al. [23]. Enzyme activity was expressed as μ moles of pyrroline 5-carboxylate formed per hour and per mg of proteins. ProDH was assayed by following the $NADP^+$ reduction at 340 nm in a 0.15 M $Na_2CO_3$-HCl buffer (pH 10.3) containing 15 mM L-proline and 1.5 mM $NADP^+$ [24].

## Results

*Changes in plant response to osmotic stress in relation to nitrogen availability*

Plants subjected to limited nitrogen supply showed an inhibition of their growth, which amounted to approximately 70% of control (at the end of the treatment) (Fig. 1). The difference between these two treatments appeared only after 35 days. The transfer of the plants from (–N) to (+N) medium restored the growth (plants -+N). In plants with appropriate nitrogen nutrition, the leaf proline concentration regularly increased to a maximum value at 28 days. During the four last weeks of treatment, proline concentration decreased significantly. In (–N) plants, proline concentration remained low representing only 50% of the control at the end of treatment. The transfer of plants from (–N) to (+N) medium quickly restored leaf proline concentration, with a transient overshoot exceeding the proline level in control plants (Fig. 2). δ-OAT activity was poorly variable in control (Fig. 3). During the five last weeks of treatment, plants (–N) showed a high δ-OAT activity relatively to

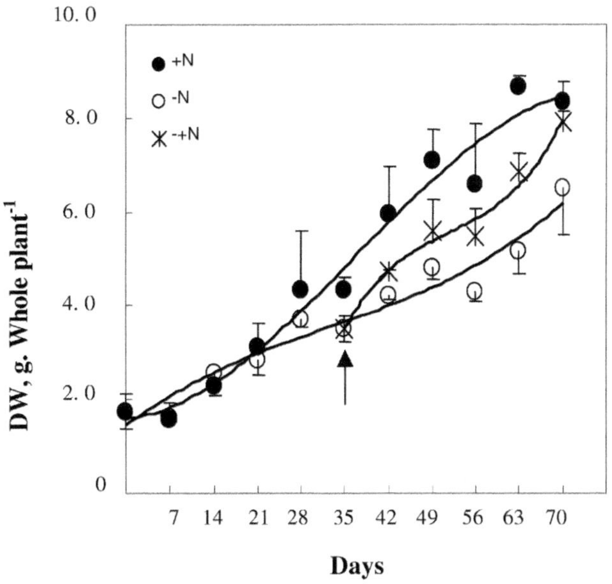


**Figure 1.** Nitrogen availability effect on whole plant dry matter production (Mean ± S.E., $n = 5$). Plants were grown on nutrient solution added with 400 mM NaCl. The arrow indicates the transfer of the plants previously subjected to nitrogen deficiency (0.28 mM) to appropriate N nutrition (14.4 mM).

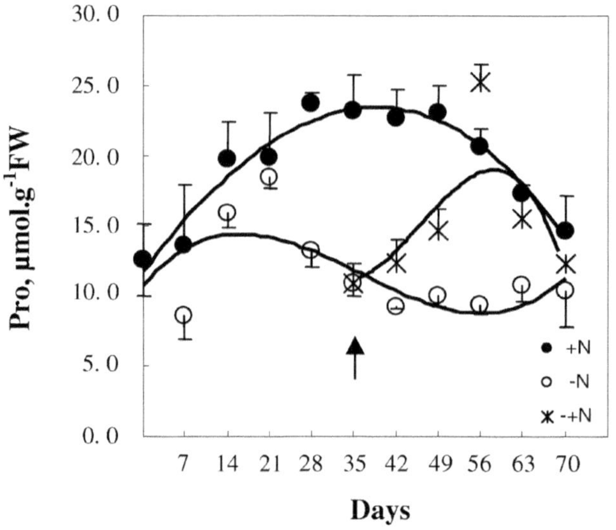


**Figure 2.** Nitrogen availability effect on proline concentration in leaves (Mean ± S.E., $n = 3$). Plants were grown on nutrient solution added with 400 mM NaCl. The arrow indicates the transfer of the plants previously subjected to nitrogen deficiency (0.28 mM) to appropriate N nutrition (14.4 mM).

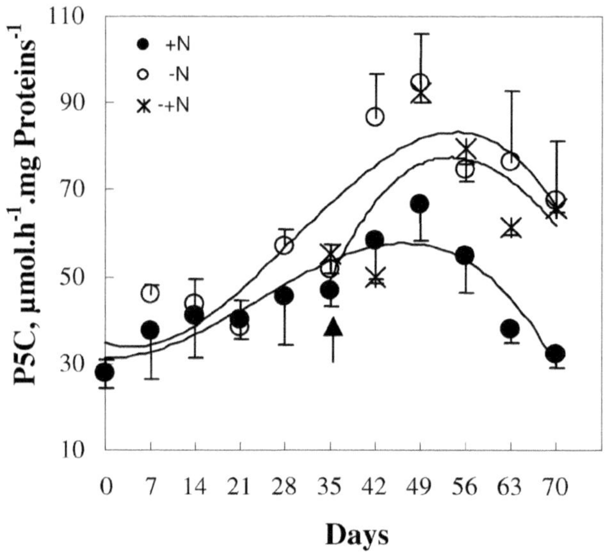


**Figure 3.** Changes in δ-OAT activity (µmol Pyrroline-5-Carboxylate (P5C) $h^{-1}$ $mg^{-1}$ protein) with nitrogen availability. Plants were grown on nutrient solution added with 400 mM NaCl. The arrow indicates the transfer of the plants previously subjected to nitrogen deficiency (0.28 mM) to appropriate N nutrition (14.4 mM).

that of control plants. The highest values of δ-OAT activity were concomitant with the lowest contents of proline. In plants supplied with N, an increase δ-OAT activity was associated with high proline concentrations. The ProDH activity presented a complex, peaking two fold in the first 3 weeks, then after 2 months of culture.

*Water stress*

The depressive effect of water deficit on the whole plant biomass appeared after 8 days of treatment (Fig. 4). After 40 days of water stress, the dry matter production of dehydrated plants was 44 % of the control. Rehydrating the plants after 28 day-long water stress allowed growth recovery, although the dry weight of the rewatered plants did not reach the level of the control. In control plants, proline accumulation remained almost unchanged during the whole period of treatment (about 10 μ mol. $g^{-1}$ FW) (Fig. 5). Water deficit induced an increase in proline levels which grew with time. At the end of the experiment, plants submitted to water deficit accumulated twice more proline than control. Proline concentration in rehydrated plants decreased quickly, to values close the control. Contrasting behavior was revealed for ProDH protein, which was high in control, low in stressed plants, and high in rewatered plants.

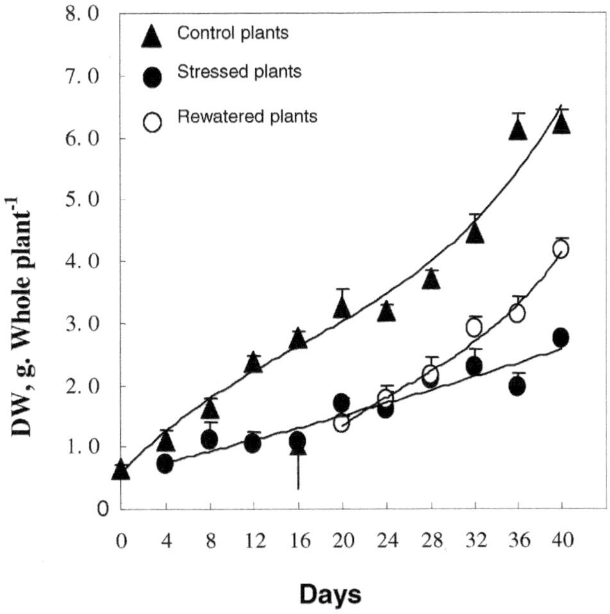


**Figure 4.** Water deficit stress effect on whole plant dry matter production (Mean ± S.E., $n = 3$). Controlled plants: plants grown on 100 % field capacity (FC), Stressed plants: plants grown on 25 % FC, rewatered plants: plants cultivated during 16 days on 25 % FC, and then transferred on 100 % FC. The arrow indicates the rewatering of the plants previously subjected to water deficit.

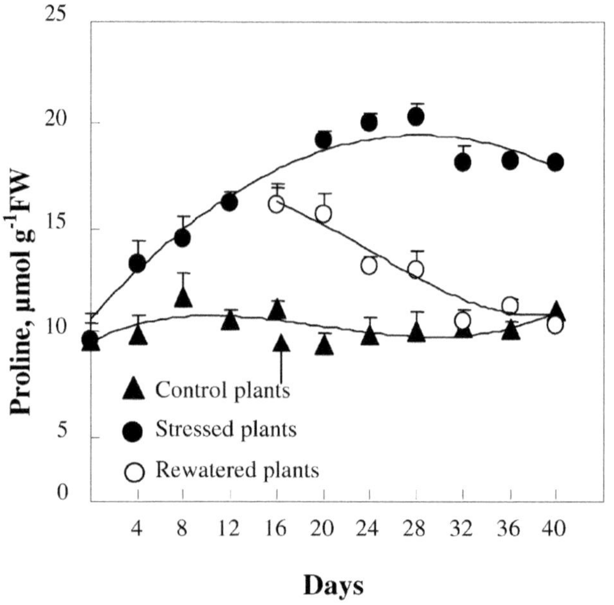


**Figure 5.** Changes in of *S. portulacastrum* leaf proline content (μmol $g^{-1}$ FW) with water availability in the culture substrate (Mean ± S.E., $n = 5$). The arrow indicates the rewatering of the plants previously subjected to water deficit.

## Discussion

Our results show that plants subjected to 400 mM NaCl (plants +N) expressed the same potentialities of growth as those cultivated in absence of salt. So, salt constraint did not affect the production of biomass at *S. portulacastrum*. It involved a significant increase in the contents of proline at the plants. But this phenomenon depends on the availability of the nitrogen. Indeed, at the plants submitted to a nitrogen deficiency as well as the growth, the accumulation of the proline was limited compared to (+N) plants. The water deficit reduced considerably growth without leading to visual toxicity symptoms (chlorosis or necrosis). It induced also a significant increase in proline concentration which reached 20–25 μmol.g $^{-1}$ FW. These results indicate that the plant reacts to the salt and water constraints by an accumulation of proline, and they suggest that the availability of nitrogen was essential to this response. To evaluate whether the accumulation of proline is an active process, we measured the activity of δ-OAT and ProDH, enzymes involved respectively in proline biosynthesis and catabolism. The δ-OAT activity increased under stress. However, in N deficient plants we showed also an increase in δ-OAT activity concomitant with a decrease in proline concentrations. These data suggest that this enzyme, normally involved in proline biosynthesis, can be also implied in its catabolism. This behavior was observed only in the animal cells. Indeed, δ-OAT interconverts P5C into ornithine and, therefore plays an important role in both synthesis and degradation of proline [25]. During water stress, an inhibition of ProDH activity was concomitant with the absence of the protein band of ProDH. The increase of ProDH activity and the higher intensity of the protein band of this enzyme observed in plants subjected simultaneously to salt stress and N deficiency suggest that the nitrogen supply eliminate the inhibitory effect of salt on this enzyme. ProDH is normally induced by proline. However, Peng et al. [26] showed that this induction doses not occur under osmotic stress. This is in agreement with our result obtained in *S. portulacastrum* submitted to water stress. But under salt stress associated to N deficiency, an over-expression of ProDH was showed. According to Ahmed and Hellebust [27], 90 % of carbon and nitrogen in the soluble compounds are represented by amino acids. The degradation of the proline in glutamate is, thus, a potential source of energy.

## References

1 Skriver K, Mundy J (1990) Gene expression in response to abscisic acid and osmotic stress. *Plant Cell* 2: 503–512
2 Munns R (2002) Comparative physiology of salt and water stress. *Plant Cell Environ* 25: 239–250
3 Zhu JK (2001) Plant salt tolerance. *Trends Plant Sci* 6: 66–71
4 Zhang H, Forde BG (1998) An *Arabidopsis* MADS box gene that controls nutrient-induced changes in root architecture. *Science* 279: 407–409
5 Wilson C, Lesch SM, Grieve CM (2000) Growth stage modulates salinity tolerance of New Zealand spinach (*Tetragonia tetragonioides*, Pall.) and Red Orach (*Atriplex hortensis* L.). *Ann. Bot* 85: 501–509
6 Messedi D, Labidi N, Grignon C, Abdelly C (2004) Limits imposed by salt to the growth of the halophyte *Sesuvium portulacastrum*. *J Plant Nutr Soil Sci* 167: 720–725
7 Sakamoto W, Tamura T, Hanba-Tomita Y, Sodmergen and Murata M (2002) The *VAR1* locus of *Arabidopsis* encodes a chloroplastic FtsH and is responsible for leaf variegation in the mutant alleles. *Genes to Cells* 7: 769–780
8 Rhodes D, Hanson AD (1993) Quaternary ammonium and tertiary sulfonium compounds in higher plants. *Annu Rev Plant Physiol* 44: 357–384
9 Kumara SG, Reddy AM, Sudhakar C (2003) NaCl effects on proline metabolism in tow high yielding genotypes of mulberry (*Morus alba* L.) with contrasting salt tolerance. *Plant Science* 165: 1245–1251
10 Delauney AJ, Hu CAA, Kishor PBK, Verma DPS (1993) Cloning of ornithine-aminotransferase cDNA from *Vigna aconitifolia* by transcomplementation in *Escherichia coli* and regulation of proline biosynthesis. *J Biol Chem* 268: 18673–18678
11 Rhodes D, Handa S, Bressan RA (1986) Metabolic changes associated with adaptation of plant cells to water stress. *Plant Physiol* 82: 890–903
12 Voetberg GS, Sharp RE (1991) Growth of the maize primary root at low water potentials. Role of increased proline deposition in osmotic adjustment. *Plant Physiol* 96: 1125–1130
13 Schwacke R, Grallath S, Breitkreuz KE, Stransky E, Stransky H (1999) LeProT1, a transporter for proline, glycine betaine, and gamma-amino butyric acid in tomato pollen. *Plant Cell* 11: 377–391

14 Kiyosue T, Yoshida Y, Yamaguchi-Shinozaki K, Shinozaki K (1996) A nuclear gene encoding mitochondrial proline dehydrogenase, an enzyme involved in proline metabolism, is upregulated by proline but downregulated by dehydration in *Arabidopsis*. *Plant Cell* 8: 1323–1335
15 Verbruggen N, Hua XJ, May M, Montagu MV (1996) Environmental and developmental signals modulate proline homeostasis: evidence for a negative transcriptional regulator. *Proc Natl Acad Sci USA* 93: 8787–8791
16 Forlani G, Scainelli D, Nielsen E (1997) Delta$^1$-pyrroline-5-carboxylate dehydrogenase from cultured cells of potato. *Plant Physiol* 113: 1413–1418
17 Hewitt EJ (1966) Sand and water culture methods used in the study of plant nutrition. 2nd Ed., *Commonwealth Bureau of Horticulture tech.* Com. 22
18 Jacobson L (1951) Maintenance of iron supply in nutrient solutions by a single addition of ferric-potassium-ethylene-diamine-tetraacetate. *Plant Physiol* 26: 411–413
19 Arnon DI, Hoagland DR (1940) Crop production in artificial solution and in soils with special reference to factors affecting yields and absorption of inorganic nutrients. *Soil Sci* 50: 463–484
20 Khan MA, Ungar IA, Showalter AM (2000) The effect of salinity on the growth, water status, and ion content of a leaf succulent perennial halophyte, *Suaeda fruticosa* (L.) Forssk. *J Arid Environ* 45: 73–84
21 Bates LS, Waldren RP, Teare ID (1973) Rapid determination of free proline for water-stress studies. *Plant Soil* 39: 205–207
22 Lutts S, Majerus V, Kinet JM (1999) NaCl effects on proline metabolism in rice (*Oryza sativa*) seedlings. *Physiol Planta* 105: 450–458
23 Kim HR, Rho HW, Park BH, Kim JS, Lee MW (1994) Assay of ornithine aminotransferase with ninhydrine. *Biochemistry* 223: 205–207
24 Ruiz JM, Sanchez E, Garcia PC, Lopez-Lefebre LR, Rivero RM, Romero L (2002) Proline metabolism and NAD Kinase activity in greenbean plants subjected to cold-shock. *Phytochemistry* 59: 473–478
25 Valle D, Simell O (1995) The hyperornithinemias. In: CR Scriver, AL Beaudet, WS Sly and D Valle (Eds): *The Metabolic and Molecular Bases of Inherited Disease*, 7th ed., vol. 1. McGraw-Hill, New York, 1147–1185
26 Peng Z, Lu Q, Verma DPS (1996) Reciprocal regulation of $\Delta^1$-pyrroline-5-carboxylate synthetase and proline dehydrogenase genes controls proline levels during and after osmotic stress in plants. *Mol Gen Genet* 253: 334–341
27 Ahmad I, Hellebust JA (1984) Osmoregulation in the extremely euryhaline marine microalga *Chorella autotrophica*. *Plant Physiol* 74: 1010–1015

Biosaline Agriculture and Salinity Tolerance in Plants
Edited by M. Öztürk, Y. Waisel, M.A. Khan and G. Görk


# Linkage studies of structure, isoenzymatic diversity and some biotechnological procedures for *Salsola* species under desert saline environments

Kristina N. Toderich[1], Victoria V. Li[1], Clanton C. Black[2], Temur R. Yunusov[1], Elena V. Shuiskay[1], Gulnara K. Mardonova[1] and Lilya G. Gismatullina[1]

[1] *Department of Desert Ecology and Water Resources Research, Samarkand Branch of Academy of Sciences, 3 Timur Malik Str., Samarkand 703000, Uzbekistan*
[2] *Department of Biochemistry and Molecular Biology, University of Georgia, Life Sciences Building, Athens, GA 30602-7229, USA*

## Introduction

One of the main undesirable consequences of the process of desertification in the Central Asia countries is an amplification of salinity process resulting in a wide development of saline soils. The amplification of the salinity process under conditions of aridization of climate is caused by the high maintenance of salts both in the surface and in subsoil waters of dry lands of the Aral Sea Basin [1]. Besides that, the recent overuse of Amudarya, Zerafshan and Syrdarya river water has resulted in the waterlogging and secondary salinization (human caused) salt/affected lands on whole adjacent territories. Effects of these impacts include alteration or destruction of vegetation, frequent disappearance of many useful, endemic and relict species of desert plants, and consequently, establishment of annual plant communities dominated by weeds and exotic species. Up to 15,000 ha of pastures are annually affected by salinity and waterlogging that resulted in the reduction of population of more than 1,500 species of mammals, birds and plants in the whole of the Central Asia region [2, 3].

Today, large areas of the southern sandy Kyzylkum desert ecosystems have been negatively affected by intensive urbanization, overgrazing by livestock, cutting shrubs on fuel, cotton monoagriculture, soils and air pollution. Such type of desert human transformed ecosystems is frequently dominated by leaf-succulent plants like species of genera *Salsola*, *Climacoptera*, *Gamanthus*, *Halocharis* and many other wild native halophytes. It was suggested that harsh desert environmental factors, especially high salinity of soils, limit plant growth and the production of viable seeds. The lowest sexual reproduction ability, seed germination and, consequently, insignificant seedling survival under natural desert condition were noted for Asiatic haloxerophytes species [4–7].

We suggest that the initial exploration of natural plant cellular mechanisms affecting the phytoremediation of elemental and/or organic pollutants is quite promising for the use of desert plants in large-scale environmental clean-up efforts. Native plants have the advantage of being highly adaptive to the local climatic and edaphically contaminated and salt/affected environments.

Our previous data on the effect of desert stress factors on microscopic and submicroscopic structure of floral organs show that dryland plants and species of genera *Salsola* in particular, provide an excellent model for analysis of natural salt stress impact on the plant cellular characteristics [8–11]. However, clearly distinguishing multitudes of populations and evident polymorphism of all organs of arid species still remains very difficult to be managed, as adequate traits are lacking. High anatomical and morphological variability for many halophytes are often taken as plant adaptation to harsh desert environments that increase their chances to endure the stress imposed by salinity [7, 10, 12]. However, it is not clear to what extent this variability represents genetic, developmental or influence of environmental variation. Furthermore, little is known about variability of embryological features and the specificity of their reproductive strategies in response to arid salt stress factors. Also, there is not enough experimental data describing the procedures of plant micropropagation and callus induction for desert plants, in particular for representatives of Chenopodiaceae. Some research on *vitro* tissue culture technology was conducted on species of genera *Atriplex*, *Kochia* and *Suaeda* [13, 14]. The biotechnological procedures for *Salsola* arid species are still unknown. The main objectives of the present paper are to examine some natural mechanisms of plant salt tolerance on morphological and genetic levels, biotechnological experiments coming to the point of practical use and theoretical knowledge concerning plant resistance and tolerance to salt stress.

It was expected that the study of reproductive strategies of plants and selection of superior genotypes from natural deserts populations coupled with clonal propagation through tissue culture may offer an alternative way for improvement and conservation of arid salt/affected lands in Uzbekistan.

## Materials and methods

Various populations of annual chenopods, *Salsola lanata Pall.* (*Climacoptera lanata*), *S. praecox Litv.* and *S. pestifer*, *A. Nelson* (*S. iberica* (*Sennen et Pau*) *Botsch.*), growing under foothills semidesert and sandy desert environments (South-Eastern Kyzylkum) were analyzed. The flower organs of four woody-shrubs species of genus *Salsola* were vacuum-coated with gold and analyzed by JEOL JSM-T330 scanning electron microscope (SEM).

Variability of eight enzymatic systems, 6PGD (E.C. 1.1.1.44), MDH (E.C. 1.1.1.37), GDH (E.C. 1.4.1.2), G6PD (E.C. 1.1.1.49), PGM (E.C. 5.4.2.2), PGI (E.C. 5.3.1.9), GOT (E.C. 2.6.1.1), DIA (E.C. 1.6.99) on the basis of starch gel electrophoresis of isoenzymes from randomly chosen embryos (30–35 for each population) was studied. Enzymes were extracted by homogenization of single embryos in 80 ml of extraction buffer EDTA, KCl, $MgCl_2$, TRITON, PVP, TRIS-HCl. Isoenzymes were separated in 10 % starch gel using two buffer systems. Staining of particular enzymes and genetic interpretation of the results followed the standard techniques [15–17].

In our scientific work we used technology of plant micropropagation by seeds. The surface-sterilized embryos were placed in Petri dishes (6 cm diameter) on two types of solid medium. One of them consisted of only agar (8 g/l), another medium contained macro and microelements and vitamins according to Murashige and Skoog – MS medium; sucrose (20 g/l), agar (8 g/l) and kinetin (0.5 mg/l). Their pH was adjusted to 5.5 before autoclaving. Embryos were grown in dark condition at 24 $\pm 1°$ C. The seedlings were transferred on two types of micropropagation MS-medium, one of which consisted of IAA (0.05 mg/l) and BAP (0.1 mg/l), but another one – IAA (0.05 mg/l) and 2iP (0.01 mg/l). The culture was maintained in the air-conditioned culture room at 24 $\pm$ 1° C, 70 % relative humidity, with 16 h light/8 h dark photoperiod and 30 $\mu$mol·m$^{-2}$·s$^{-1}$ intensity. In order to induce callus, the leaf and shoot explants were placed onto various callus induction MS-media consisting of different concentrations of plant growth regulators: BAP (2.0 and 5.0 mg/l), kinetin (2.0 mg/l), 2iP (1.0 mg/l), IBA (1.0 mg/l) and 2,4-D (2.0 mg/l) in dark conditions at 25 $\pm$ 1° C for 6 weeks. Then these callus cultures were transferred onto MS-medium supplemented with 2.0 mg/l 2,4-D to get a stable viable callus culture. In order to obtain plant regeneration, callus culture were transferred onto MS-medium consisted of cytokinin BAP (2.0 mg/l) and placed at 24 $\pm$ 1° C with 16 h light/8 h dark photoperiod. All analyses have been conducted at the Department of Botany of Adam Mickiewicz University in Poznan, Poland.

## Results and discussion

The species studied by us are widespread haloxerophytes growing in open communities of inland saltmarshes and/or marshy-steppe communities of Aral-Caspian regions. All of them are considered useful for rehabilitation of arid/semiarid degraded and salt/affected lands. They possess superficial root and can be established from direct seeding and also are capable of self-sowing. All examined taxa had a low rate of field seed germination and were more tolerant to higher soils salinity at early seedling and growing stages [8, 18]. Seed reproduction can be by self-pollinating and out-crossing. *S. lanata*, forming a distinct and almost monospecific dominant plant community on desert salt/affected lands, showed a highly competitive vigor to such environments. *S. pestifer* and *S. praecox* with slender leaves, occur on dry, sandy, mostly in moderate saline soils. *S. praecox* mostly grows on sandy solonchaks of Iran-Turanian and Transcaspian lowlands. Its native wild spread populations are often small and isolated. Furthermore *S. pestifer* and *S. praecox* taxonomically are so closely related that sometimes it's difficult to distinguish their ecoforms. Both species possess both high vegetative and reproductive organ polymorphism.

### *Genetic and cytological diversity*

Our experimental data shows that the examined enzyme systems, particularly 6PGD, DIA, G6PD, GDH, GOT, MDH, PGM, and PGI, were coded by different number of loci: 14 in *S. lanata*, 17 in *S. pestifer* and 16 in *S. praecox*. The enzyme PGI is coded by three loci only in *S. pestifer* populations and two loci of G-6PD was founded in *S. lanata* populations. Moreover, in *S. lanata*, the enzymes GDH, GOT, PGI and PGM are coded by different numbers of loci than for *S. pestifer* and *S.praecox*. The general pattern of isoenzymatic bands of *S. pestifer* and *S. praecox* were similar for all loci, except PGI A, which was not detected for *S. praecox*. The proportion of polymorphic loci (0.95 criterions) was rather low: 5.9 % for *S. pestifer*, 7.1 % for *S. lanata* and 12.5 % for *S. praecox*. The mean number of alleles and genotypes per locus was not high either: from 1.1 in *S. pestifer* and *S. praecox* to 1.2 in *S. lanata* for both parameters. Generally, we found 18 alleles and 18 genotypes for *S. pestifer* and *S. praecox*, and 17 alleles and 7 genotypes for *S. lanata* in all loci of the examined enzyme systems. *S. lanata* and *S. pestifer* have similar values of genotype polymorphism index, Pg (0.025 and 0.029, respectively), as opposed to *S. praecox*, which has the highest genotype polymorphism (0.054). Values of genetic distance based on allele DN and genotype DH frequencies are summarized in Table 1. In both cases the smaller genetic distance was found for the pair *S. lanata* – *S. pestifer* and the largest for *S. pestifer* – *S. praecox*.

**Table 1.** Variability of genetic parameters of Asiatic *Salsola* species

| Populations | % of loci polymorphic* | A/L | G/L | $H_e$ | $H_o$ | F | $P_g$ |
|---|---|---|---|---|---|---|---|
| *S. lanata* | 7.1 | 1.2 | 1.2 | 0.019 | 0.031 | −0.675 | 0.025 |
| *S. pestifer* | 5.9 | 1.1 | 1.1 | 0.023 | 0.129 | −4.733 | 0.029 |
| *S. praecox* | 12.5 | 1.1 | 1.1 | 0.033 | 0.159 | −3.768 | 0.054 |

Table 1 gives the main genetic parameters calculated for the studied species: A/L-mean number of alleles per locus, G/L-mean number of genotypes per locus, He-expected heterozygosity, Ho-observed heterozygosity, F-Wright's fixation index, Pg-genotype polymorphism index.

Genetic similarities between examined species are surprisingly high: 0.996–0.999 according to Nei (IN) and 0.988–0.999 according to Hendrick (IH) [19], especially in comparison to values given by Crawford [20] for *Chenopodium* spp (0.35–0.97). Genetic similarities calculated for *S. lanata*, *S. pestifer* and *S. praecox* are as high as for conspecific populations of several *Chenopodium* species [20]. Genetic distances between the studied taxa are much lower (0.0007–0.0029) than the values estimated by Nei [21] between species (0.1–0.2) and between subspecies (0.02–0.1). Such a high isozyme monomorphism of *Salsola* species may be due to very narrow saline environmental conditions, under which only highly specialized organisms can survive [22]. This conclusion is in concordance with the hypothesis of Hamrick [19]. The authors suggest that strong directional selection might result in lower levels of genetic variation in arid plant populations. According to Golding [23] environment factors also shape frequencies of alleles, thus taxa experiencing similar conditions may exhibit similar allele frequencies. The above genetic data suggest that a significant component of the great morphological variation of the examined taxa may result from phenotypic plasticity. By contrast, Rilke and Reimann [12] present

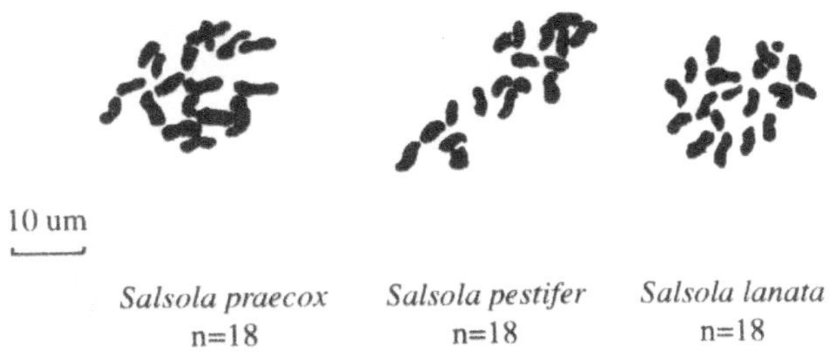


**Figure 1.** Karyotype characteristics for some Asiatic annual species of genus *Salsola*

that differential saline tolerance of *Salsola kali* from northern Europe suggests some genetic differentiation.

Thus, our results show rather low levels of isoenzymatic and genetic variation of investigated populations of annual *Salsola* species. Nevertheless, this study demonstrates the usefulness of electrophoretic technique in exploring genetic diversity and taxonomic position of many close related species of the Chenopodiaceae family. Obviously, our findings require further studies on wider plant material in respect to genetic variation and its relation to environmental conditions.

The preliminarily karyological analysis revealed that all three examined species have the same number of chromosomes: 2n = 18 (Fig. 1). However, further investigations are necessary.

*Variability of reproductive organs and its adaptive trends*

By means of scanning electron microscopy (SEM) analysis it was defined that fruit tepals morphology and anatomy plays a significant role in the species delimitation. Besides this, many fruits' morphological and structural traits are indicated to the adaptive level of *Salsola* species to deserts saline environments that in some way determine their seed viability and germination ability. The studied annual C4-*Salsola* species showed an evident polymorphism in the morphology, density and sizes of papillae, hairs, trichomes and salt glands (shape of their head: mainly clavate or capitate). Abundant papillae prickle hairs and secreted salts on the ridges of adaxial bracts/bracteoles are described (Fig. 2). Frequently, salt glands are globose or club-shaped and readily distinguishable from unicellular papillae and sharp-pointed prickles. This is a prominent mechanism of salts accumulation both on the surface and/or internal cellular tissue (epidermis and parenchyma) in the form of crystals (usually containing oxalates), which are often secreted chiefly in the form of clusters both in solution, sands or others types are marked. The accumulation of toxic salts into the reproductive organs, embryo tissues in particular, has not been detected. It was also found that the species of *Climacoptera* develops mostly unicellular non-glandular trichomes (Fig. 3) or prickle hairs possessing a smooth (warted) succulent surface, while the dry/sclerified species (*S. pestifer* and *S. praecox*) have an indulating epidermal surface with a tall adaxial ridges alternating with deep grooves. The salt glandular structures and salt secretion on the bract/bracteoles of these species

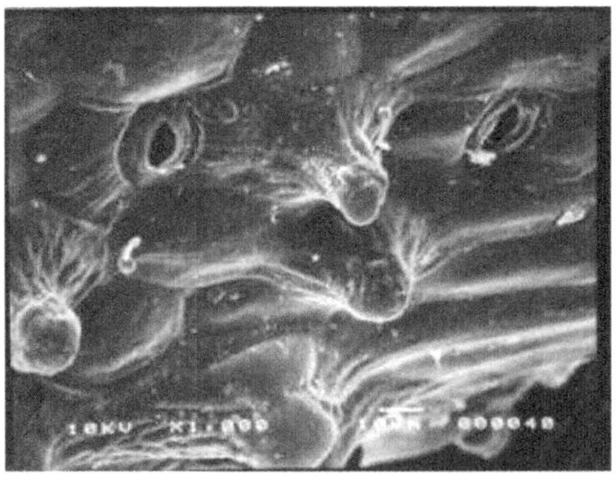

**Figure 2.** The morphology of vesicular and short peltate trichomes with sunken stomata on bracts in *S.pestifer* (Buchara ecotype)

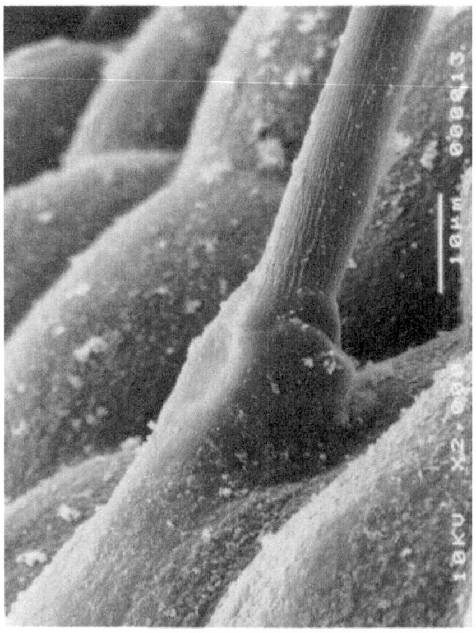

**Figure 3.** SEM micrograph showing surface features and morphology of non-glandular, unicellular hair of bracts in *Climacoptera lanata* (Buchara ecotype)

become abundant when plants are exposed to high saline environments. We supposed that these parameters could be used as discriminating characters between different ecological halophytes *Salsola* groups. Variation in the indumentum density is believed to be mainly the effect of stress/desert environmental factors and/or even herbivores pressure [24].

A lyzicarpous utricle, winged and monospermous fruit are characterized for all studied species. The seeds are small, dark-brown, orbicular and horizontally arranged. The embryo is large and spirally coiled. A strong exogenic type of dormancy was found for these annual Asiatic *Salsola* species. This type of seed dormancy is closely related to the 'perianth segments' structure and its chemical composition presents some physical barriers and chemical inhibitors (phenolic compounds, abcisic and nicotinic acids, surplus of mineral salts), reducing the water and gaseous exchange to the embryo [25]. Seed loses their viability 6–8 months after harvesting. Long-term stratification, 1.0–2.5 months under 2–4° C/15–28° C night/day cycle and/or seed's wash (2–3 times/10 min) with distilled water from inhibitors positively contribute to the seed germination. The destruction of intactness of seed coat or removal of fruit cover is also effective to increase seed germination. Dry storage (under 20/28°C in special hermetic cameras) of seeds gives positive results to increase and/or keep its germinability both in laboratory and field conditions.

*Plant micropropagation and callus induction procedures*

We found that seeds of all examined species germinate within 1–2 days. The rate of seed germination was higher on pure agar than on MS medium with kinetin as a growth regulator. It was more peculiar for *S. pestifer* (sand desert population). Shoot proliferation from seeds was noticed on both used micropropagation mediums. As is known, different parts of a plant such as roots, annual and perennial shoots, leaves, flowers, and ovary have been used as explants to obtain callus culture [26–29]. In our experiments we have used fragments of shoots and leaves to induce callus. It was determined that the growth regulators in MS medium have a different influence on callus induction even within limits of the same species. As shown in Table 2, callus initiation of *S. pestifer* (foothills semidesert population) was much evident on the medium supplemented by BAP (2.0 mg/l); while IBA (1.0 mg/l) was the most suitable growth substance that induces callus culture of *S. pestifer* (sand desert population).

Process of callus induction depended also on explant types. It was determined that callus culture for both populations of *S. pestifer* showed the best development from shoot explants. At the same time the leaf explants of *S. lanata* induced callus culture more often when comparing with shoot explants. It is important to note that callus formation started within 1 week of inoculation on MS medium supplemented by BAP (2.0 mg/l) as a growth regulator for all of the studied plants. The earliest sign of callus formation from cuttings explants of all species has been visible within 1 week on MS medium supplemented by BAP (2.0 mg/l). This medium was also the most suitable medium for callus induction from both shoot and leaf explants for foothills semidesert populations of *S. pestifer*. MS medium with IBA, however, seemed to be one of the optimal medium for callus culture induction in the case of the sandy desert populations of *S. pestifer* (Tab. 2).

According to some authors, the auxine 2,4-D is most frequently used to initiate plant callus culture [28, 30]. For our purpose, 2,4-D appeared to be the most suitable growth regulator to induce callus culture of *S. lanata* and to obtain the stable viable callus for all three investigated species. The explants of *S. pestifer* produced two type of callus: a white, transparent, soft and white, compact. For *S. lanata*, a yellowish, compact callus culture was observed.

The plant regeneration both from lateral buds and callus culture of *S. lanata* was obtained by using MS medium on 2.0 mg/l BAP.

**Table 2.** Callus induction from shoot and leaf explants on different types of medium

| MS medium with growth regulators (mg/l) | Shoot explants | | Leaf explants | |
|---|---|---|---|---|
| | Total, number | With callus induction (%) | Total, number | With callus induction (%) |
| *Salsola pestifer* (foothills semidesert population) | | | | |
| BAP (2.0) | 46 | 100.0 | 77 | 81.8 |
| BAP (5.0) | 7 | 42.9 | 46 | 15.2 |
| Kin (2.0) | 20 | 40.0 | 20 | 65.0 |
| 2iP (1.0) | 19 | – | 20 | – |
| IBA (1.0) | 20 | 95.0 | 20 | 35.0 |
| 2,4-D (2.0) | 13 | – | 13 | – |
| *Salsola pestifer* (sand desert population) | | | | |
| BAP (2.0) | 48 | 91.7 | 90 | 68.9 |
| BAP (5.0) | 2 | – | 49 | – |
| Kin (2.0) | 19 | 94.7 | 16 | 100.0 |
| 2iP (1.0) | 18 | 100.0 | 14 | 28.6 |
| IBA (1.0) | 25 | 100.0 | 21 | 100.0 |
| 2,4-D (2.0) | 18 | 94.4 | 20 | 100.0 |
| *Salsola lanata* | | | | |
| BAP (2.0) | 58 | 5.2 | 84 | 38.1 |
| BAP (5.0) | 11 | – | 21 | 4.8 |
| Kin (2.0) | 18 | – | 17 | 64.7 |
| 2iP (1.0) | 23 | – | 20 | 80.0 |
| IBA (1.0) | 19 | 31.6 | 19 | 31.6 |
| 2,4-D (2.0) | 18 | 100.0 | 15 | 93.3 |

## Conclusions

Studies on reproductive organs' structural analysis and genetic variation of three Asiatic haloxerophytes of the genus *Salsola* demonstrate the usefulness of SEM and electrophoretic techniques in exploring genetic diversity and taxonomic position of many close related species of the genus *Salsola* from the Chenopodiaceae family. Some structural traits of salt glands/hairs morphology could be used as discriminating characters between different ecological *Salsola* halophytes groups.

Preliminarily methodological work on plant micropropagation and callus induction showed the potential utilization of Asiatic annual haloxerophytes in selection of salt-tolerant genotypes for dealing with rehabilitation of saline affected lands in the continental desert climate of Uzbekistan.

Thus, our results should enable more rigorous selection of salt/drought tolerant plants for rehabilitation of saline/sodic sandy desert lands in Uzbekistan. The received data can be integrated in a global database to draw a full description of the genetic variation of desert plant resources on a Central Asian scale. It is an important step towards practical and accurate methods for the determination of the key problem in a modern rangelands management in Uzbekistan.

*Acknowledgements*
This work was conducted at the Department of Botany, Institute of Experimental Biology, Adam Miczkiewicz University in Poland, under leadership of Professor Adam Wozny. Work was partially funded by Kasa Mianowskiego Foundation for Scientific Promotion in Warsaw (Poland).

## References

1 Tsvetsinskaya EA, Vainberg BI, Glushko EV (2002) An integrated assessment of landscape evolution, long-term climate variability, and land use in the Amudarya Prisarykamysh delta. *J Arid Environments* 51: 363–381
2 National strategy and action plan on conservation and sustainable use of biological diversity in the Republic of Kazakhstan (1999) – a report of Ministry of Natural Resources and Protection of Environment of the Republic of Kazakstan: 102
3 De Beurs KM, Henebry GM (2004) Land surface phenology, climatic variation, and institutional change: analyzing agricultural land cover change in Kazakhstan. *Remote Sensing of Environment* 89: 497–509
4 Nazariyuk LA (1968) Biological characteristics of some annual salsolas forage species in relation to the improvement of ranges on the adyr of Nishan steppe. (Biologicheskaya kharakteristika nekotorykh kormovykh vidov odnoletnikh solyanok v svyazi s uluchsheniem pastbishch na adyrakh Nishanskoi stepi.) Avtoreferat dissertatsii na soiskanie uchyonoi stepeni kandidata biologicheskikh nauk, Tashkent: 18
5 Shamsutdinova EZ, Myasodoev NA, Kalinkina LG, Baburina OK, Naumova TG, Balnokin YV (1997) Impact of soil salinization on amino contents in halophyte species. *Problemy Osvoeniya Pustyn* 10(3): 70–73
6 Khan MA, Ungar IA (2000) Alleviation of innate and salinity induced dormancy in*Atriplex griffithii* Moq. var. stocksii Boiss. Department of Environmental and Plant Biology, Ohio University, Athens, OH, 45701 2979, USA. *Seed Sci and Technol* 28(1): 29–37
7 Toderich KN, Aparin VB, Tsukatanu T, Konkin AB (2003) A strategy for land rehabilitation by salt and heavy metal removal using the integration of Asiatic desert plant diversity. *Chinese J Arid Land Geography* V.26: 10: 150–159
8 Li VV (2000) Effect of various types of substrate salinity on seed germination of *Climacoptera lanata* (Pall.). Abstracts of International Seminar, 'Prospects for Saline Agriculture', Islamabad: 86
9 Li VV (2000) The opportunities of using annual *Salsola* species associated with rehabilitation of salt/affected lands in South-Eastern Kyzylkum. Abstracts of the International Scientific Conference, 'Problems of desertification in arid zones', 16–18 October 2000, Samarkand, Uzbekistan: 69 (Russian)
10 Toderich KN, Goldshtein RI, Aparin VB, Idzikowska K, Rashidova GSh (2001) Environmental state and analysis of phytogenetic resources of halophytic plants for rehabilitation and livestock feeding in arid sandy deserts of Uzbekistan. In: SW Breckle, W Weste, W Wucherer (eds): *Sustainable land use in deserts.* Springer, Berlin-Heidelberg, New York 154–165
11 Toderich KN, Tsukatani T, Petukhov OF, Gruthinov VA,Khujanazarov T, Juylova EA, (2004) Risk assessment of environmental contaminants of Asiatic Deserts Ecosystems in relation to plant distribution and structure. *J Arid Land Studies* 14S: 33–36
12 Rilke S, Reiman C (1996) Morphological and ecophysiological differences between the subspecies of *Salsola kali* L. in Europe: results of culture experiments. *Flora (Jena)* 191: 363–376

13 Kenny L, Caligari PDS (1996) Androgenesis of salt tolerant shrub *Atriplex glauca. Plant Cell Reports* 15: 829–832
14 Thiyagarajah M, Fry SC, Yeo AR (1996) *In vitro* salt tolerance of cell wall enzymes from halophytes and glycophytes. *J Exp Botany* 47 (304): 1717–1724
15 Rudin D, Ekberg I (1978) Linkage studies in *Pinus sylvestris* using macrogametophyte allozymes. *Silvae Genetica* 27: 1–12
16 Szmidt AE, Yazdani R (1984) Electrophoretic studies of genetic polymorphism of shikimate and 6-phosphogluconate dehydrogenases in Scots pine (*Pinus sylvestris* L.) in genetic studies of Scots pine domestication by means of isoenzyme analysis. PhD dissertation SUAS, Umea, Sweden: 6–13
17 Muona O, Szmidt A (1985) A multilocus study of natural populations of *Pinus sylvestris* in population genetics in forestry. *Lect Notes Biomath* 60: 226–240
18 Toderich KN, Goldshtein RI, Aparin SB, Rashidova GSh (1999) Ecological state and an analysis of halophytic resources of arid/saline deserts of Kyzylkum. Ecological problems of sustainable Land. Use in Deserts, Bonn
19 Hamrick JL, Mitton JB, Linhart YB (1981) Levels of genetic variation in trees: The influence of life history characteristics. *Proc Symp on Isozymes of N Amer Forest Trees and Forest Insects. Pacific SW For and Range Expt Stat Tech Report PSW-48: 35–41*
20 Crawford DJ (1983) Phylogenetic and systematic inferences from electroforetic studies. Part A. In: SD Tanskley, TJ Orton (eds) Isozymes in plant genetics and breeding. Elsevier Science Publishers B.V., Amsterdam 237–244
21 Nei M, Roychoudhury AK (1978) Sampling variances of heterozygosity and genetic distance. *Genetics* 76: 379–390
22 Wojnicka-Poltorak A, Chudzinska E, Shuiskaya E, Barczak H, Toderich K, Prus-Glowacki W (2002) Isoenzymatic and cytological studies of some Asiatic species of genus Salsola. *Acta Societatis Botanicorum Poloniae* 71(2): 115–120
23 Golding B (1994) Genetic relationship among 19 accessions of six species of *Chenopodium* L. by random amplified polymorphic DNA fragments (RAPD). In: PM Ruas, A Bonifacio, CF Ruas, DJ Fairbanks, WR Andersen Non-neutral evolution. Theories and molecular data. *Euphytica* 105: 25–32
24 Johnson HB (1975) Plant pubescence: an ecological perspective. *Bot Rev* 41: 233–258
25 Gintzburger G, Toderich KN, Mardonov BK, Makhmudov MM (2003) *Rangelands of arid and semiarid zones of Uzbekistan.* CIRAD-ICARDA Publisher, France 478
26 Upadhyay N, Makoveychuk AY, Nikolaeva LA, Batygina TB (1992) Organogenesis and somatic embryogenesis in leaf callus culture of *Rauwolfia caffra* Sond. *J Plant Physiol* 140: 218–222
27 Casstellar MR, Iborra JL (1997) Callus induction from explants of *Crocus sativus*. J Plant Biochem Biotechnol 6(2): 97–100
28 Luo JP, Jia JF (1998) Callus induction and plant regeneration from hypocotyl explants of the forage legume *Astragalus adsurgens. Plant Cell Reports* 17: 567–570
29 Elhaak MA (1999) Effect of abscisic acid on growth and certain osmoregulatory metabolites in the leaf callus of *Gymnocarpos decandrum. Arab Gulf J Sci Res* 17(1): 95–109
30 Zheng MY, Konzak CF (1999) Effect of 2,4-dichlorophenoxyacetic acid on callus induction and plant regeneration in anther culture of wheat (*Triticum aestivum* L.). *Plant Cell Reports* 19: 69–73

Biosaline Agriculture and Salinity Tolerance in Plants
Edited by M. Öztürk, Y. Waisel, M.A. Khan and G. Görk


# Kinetics of the antioxidant response to salinity in *Crithmum maritimum*

Nader Ben Amor[1], Karim Ben Hamed[1], Annamaria Ranieri[2] and Chedly Abdelly[1]

[1] *Laboratoire d'Adaptation des Plantes aux Stress Abiotiques, Institut National de Recherche Scientifique et Technique, BP 95, Hammam-Lif 2050, Tunisia*
[2] *Dipartimento Chimica e Biotecnologie Agrarie, Università degli studi di Pisa, Pisa, Italy*

## Introduction

Salinity limits the production of approximately 40% of the world's agricultural land [1]. In order to overcome the decline of cultivated areas and the high demands for food and energy, a particular interest was accorded to the salty lands. To re-green these areas, two strategies have been developed: i) the genetic manipulation of common crop species for increased salt tolerance, and ii) the utilization of naturally salt-tolerant species (halophytes) [2].

Mechanisms of salt tolerance are of two main types: those minimizing the entry of salt into the plant, and those preventing the salt concentration in the cytoplasm. Halophytes, naturally salt tolerant plants, express both these properties. They exclude salt well, but effectively compartmentalize in vacuoles the salt that inevitably gets in, which ultimately allows them to grow for a long period in saline soils.

The involvement of antioxidative response systems (ARS) in salt tolerance was often reported in halophytes [1, 3–5]. The primary components of this defence system include carotenoids, ascorbate, glutathione, and tocopherol as well as antioxidant enzymes such as superoxide disumutases (SOD), catalase, peroxidases and the enzymes of the ascorbate glutathione cycle (e.g., ascorbate peroxidase, monodehydroascorbate reductase, dehydroascorbate reductase and glutathione reductase) [6, 4]. Plants containing high concentrations of antioxidants show considerable resistance to the oxidative damage caused by the activated oxygen species [7, 5]. Recent reports on the response of plant antioxidant enzymes to salinity have indicated several distinct patterns, which varied according to the species and the tissue analyzed [8, 9]. The aim of this study was to investigate the changes in activities of the antioxidant enzymes, SOD, catalase (CAT, EC 1.11.1.6) and peroxidase (POD, EC 1.11.1.7) in the leaves and roots of *Crithmum maritimum*, a perennial wild halophyte, in response to increasing NaCl.

## Material and methods

Seeds of *C. maritimum* collected from rocky coasts of Tabarka (160 km north of Tunisia) were sterilized in 0.2 % (w/v) sodium hypochlorite for 3 min and germinated on filter paper in Petri dishes in a growth chamber at controlled conditions (16 h/8 h light/dark regime; 200 $\mu$mol·m$^{-2}$·s$^{-1}$ photosynthetic active radiation (PAR); 15–25° C temperature; 70–90 % relative humidity). Two-leaf seedlings were transferred to 5 l plastic pots (5 plants per pot) and were hydroponically cultivated, using aerated Hewitt nutrient solution (pH 7.3, EC 2.7 mS.cm$^{-1}$). Plants were separated in three groups irrigated with a nutrient solution supplemented with different concentrations of NaCl (0, 50 and 200 mM). To avoid osmotic shock, salt was increased gradually up to the final salt concentration. The nutrient solutions were renewed every 4 days. Intermediate harvests were carried out after 0, 10, 20, 30 and 60 days of salt treatment. Fresh samples (shoot and root) were immediately frozen in liquid nitrogen and stored at –80° C until biochemical determination.

The plant material was extracted at 4° C in 100 mM Tris-HCl buffer (pH 8.0) containing 10 mM EDTA, 50 mM KCl, 20 mM $MgCl_2$, 0.5 mM PMSF, 1 mM DTT, 0.1 % (v/v) Triton X-100, and 10 % (W/W) PVP. The ratio of plant material to buffer was 1:3. The homogenate was centrifuged at $14,000 \times g$ for 30 min at 4° C and the supernatant was used for enzymatic analysis. Three replicates per treatment were used. Protein concentration was determined using bovine serum albumin as a standard [10].

Catalase activity was performed by following the $H_2O_2$ disumutation at 240 nm in a reaction mixture (3 ml) composed of 50 mM sodium phosphate buffer (pH 7.0), 30 % $H_2O_2$ (DO 0.52–0.55 at 240 nm) [11]. CAT activity was expressed as units ($\mu$mol of $H_2O_2$ decomposed per min) per mg of protein. Peroxidase activity was determined spectrophotometrically by measuring the oxidation of o-dianisidine (3, 3'-dimethoxybenzidine) at 460 nm [12]. POD activity was expressed as units ($\mu$mol of oxidized dianisidine per min) per mg of protein.

## Results

### *$H_2O_2$-detoxifying enzyme responses in roots*

CAT activity increased 20 and 30 days after the start of NaCl treatment, respectively in controls and plants treated with 50 mM NaCl, and reached a maximal value at the end of treatment (60 days) (Fig. 1). POD activity increased gradually with time in both treated and control plants. The effect of salt stress on POD activities was only perceptible at the end of treatment (Fig. 2). On the other hand, the two enzyme (CAT and POD) activities showed variations in the 200 mM NaCl-treated plants, especially at the beginning of the treatment.

### *$H_2O_2$-detoxifying enzyme response in shoots*

CAT activity of both control and NaCl-treated plants increased slightly 10 days after the onset of the treatment (Fig. 3). Thereafter, CAT activity remained constant until

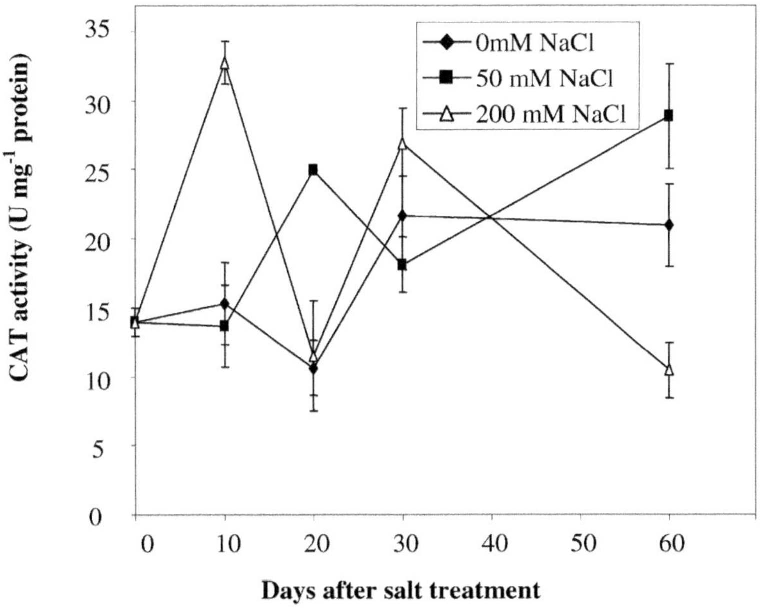


**Figure 1.** Changes in CAT activity (Unit $mg^{-1}$ protein) of *C. maritimum* roots in response to NaCl. Means of 5 replicates (± SE, P = 0.05)

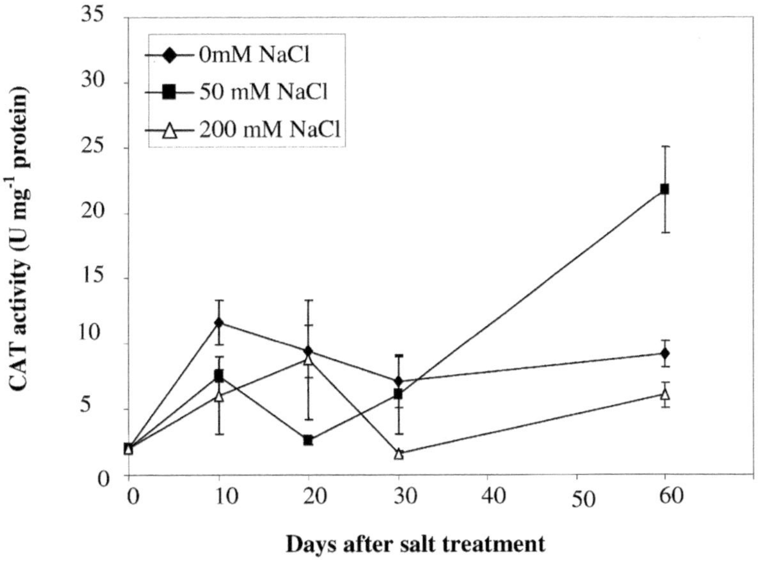


**Figure 2.** Changes in CAT activity (Unit $mg^{-1}$ protein) of *C. maritimum* shoots in response to NaCl. Means of 5 replicates (± SE, P = 0.05)

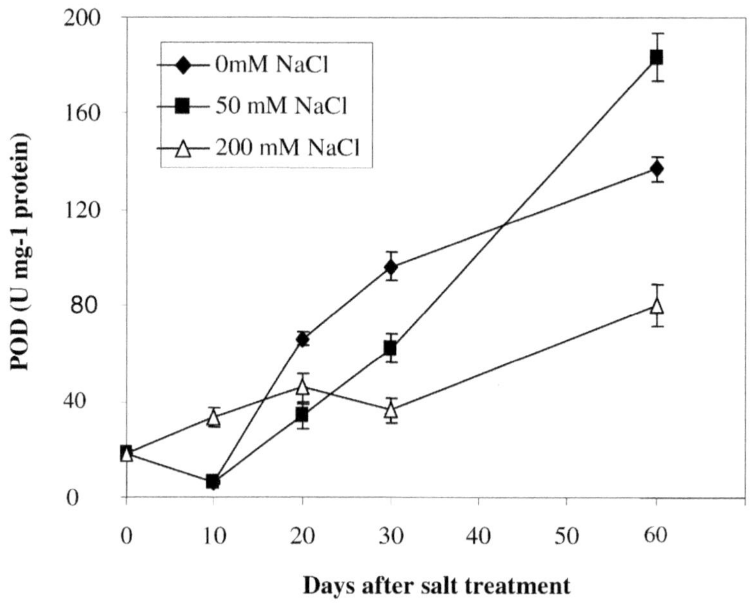


**Figure 3.** Changes in POD activity (Unit $mg^{-1}$ protein) of *C. maritimum* roots in response to NaCl. Means of 5 replicates (± SE, P = 0.05)

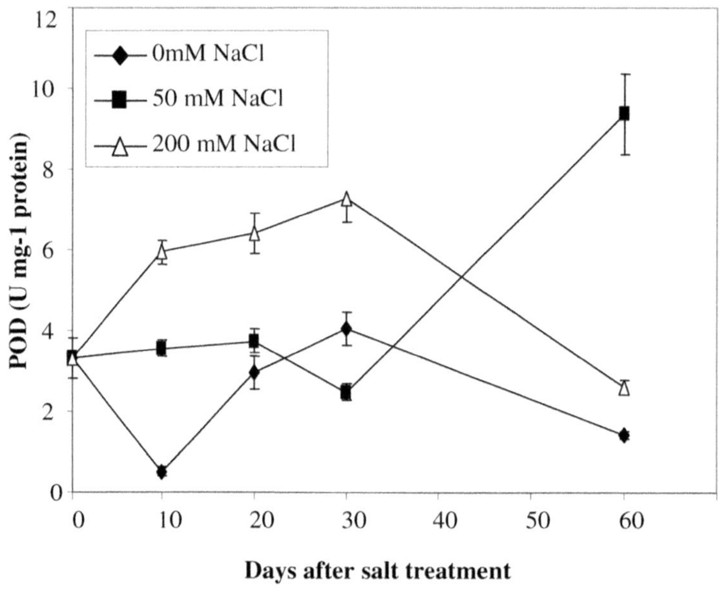


**Figure 4.** Changes in POD activity (Unit $mg^{-1}$ protein) of *C. maritimum* shoots in response to NaCl. Means of 5 replicates (± SE, P = 0.05)

the end of the treatment in both control and 200 mM NaCl-treated plants. In addition, CAT activity increased substantially the plants treated with 50 mM NaCl (Fig. 3). Changes in POD activities were time and dose-dependant. At the beginning of the treatment, the highest activity was found in plants treated with 200 mM. The most important activities at the end of salt treatment were recorded in shoot extracts from plants treated with 50 mM NaCl (Fig. 4).

## Discussion

Several mechanisms are developed by plants to detoxify the reactive oxygen species (ROS). It is clearly important to establish whether the exposure of plants to salinity causes a detrimental or stiumulatory effect on the enzymes involved in this detoxification process. Salt stress tolerance has been correlated to an improved oxidative stress response in several crops [1, 13–15, 5]. SOD, CAT, and POD are among the major antioxidant enzymes involved in scavenging ROS [16]. Previous studies investigating the impact of NaCl on CAT activity in higher plants, reported a decrease in activity, but with a variation during the time of treatment. The decline in CAT activity due to salt stress in cotton, in sunflower, and *Bruguiera parviflora* [17] agrees with our results in *C. maritimum*, when treated with 200 mM NaCl. Furthermore, similarly to *Sedum album* [18], an inductive response in CAT at low NaCl concentration (50 mM) was found in *C. maritimum.*

Unspecific peroxidase activity has been used as a general indicator of stress induced by high temperature, salinity and drought [19, 20]. A salt-induced increase of peroxidase activity was reported to occur in foxtail millet [21] and rice [22]. Similarly, our data show a high level of induction in POX activity in *C. maritimum*. The extents of increase in activity appear to be significant in relation to the duration of the treatment and the concentration of salt.

Our findings strongly support an induced increase in $H_2O_2$-scavenging enzyme activity (CAT and POD) in both leaves and roots of *C. maritimum* cultivated in the presence of 50 mM NaCl. Currently, it is not known whether the increase might be due to an upregulation of the genes controlling the synthesis of these enzymes or to an increased activation of constitutive enzyme pools. Since other studies have indicated that salt stress increases the synthesis of certain proteins in some plants [23], it is likely that the salt-induced increase in the antioxidant enzyme activities observed in *C. maritimum* might be due to changes in the antioxidant gene encoding the enzyme activities.

## References

1 Rout NP, Shaw BP (2001) Salt tolerance in aquatic macrophytes: possible involvement of the antioxidative enzymes. *Plant Sci* 160: 415–423
2 Maggio A, Reddy MP, Joly RJ (2000) Leaf gas exchange and solute accuumulation in the halophyte *Salvadora persica* grown at moderate salinity. *Environ Exp Bot* 44: 31–38
3 Tester M, Davenport R (2003) $Na^+$ tolerance and $Na^+$ transport in higher plants. *Ann Bot* 91: 503–527

4 Wang B, Luttge U, Ratajczak R (2004) Specific regulation of SOD isoforms by NaCl and osmotic stress in leaves on the $C_3$ halophyte *Suaeda salsa* L. *J Plant Physiol* 161: 285–293
5 Ashraf M, Harris PJC (2004) Potential biochemical indicators of salinity tolerance in plants. *Plant Sci* 166: 3–16
6 Asada K (1999) The water-water cycle in chloroplasts: scavenging of active oxygen and dissipation of excess photons. *Annu Rev Plant Physiol Plant Mol Biol* 50: 601–639
7 Shalata A, Tal M (1998) The effect of salt stress on lipid peroxidation and antioxidants in the leaf of the cultivated tomato and its wild salt-tolerant relative *Lycopersicon pennelli*. *Physiol Plant* 104: 169–174
8 Gossett DR, Millhollon EP, Lucas MC, Banks SW, Msrney MM (1994) The effects of NaCl on antioxidant enzyme activities in callus tissue of salt-tolerant and salt-sensitive cotton cultivars (*Gossypium hirsutum* L.). *Plant Cell Rep* 13: 498–503
9 Ben Amor N, Ben Hamed K, Debez A, Grignon C, Abdelly C (2005) Physiological and antioxidant responses of perennial halophyte *Crithmum maritimum* to salinity. *Plant Sci* 168: 889–899
10 Bradford MM (1976) A rapid and sensitive method for the quantitation of microgram quantities of protein utilizing the principle of protein-dye binding. *Anal Biochem* 72: 248–254
11 Aebi H (1984) Catalase *in vitro*. *Method Enzymol* 105: 121–126
12 Ranieri A, Petacco F, Catagna A, Soldatini GF (2000) Redox state and peroxidase system in sunflower plants exposed to ozone. *Plant Sci* 159: 159–168
13 Arbona V, Flors V, Garcia-Agustin P, Gomez-Cadenas A (2003) Enzymatic and non enzymatic antioxidant responses of *Carrizo citrange*, a salt sensitive *Citrus* rootstock, to different levels of salinity. *Plant Cell Physiol* 44: 388–394
14 Shalata A, Mittova V, Volokita M, Guy M, Tal M (2001) Response of the cultivated tomato and its wild salt-tolerant relative *Lycopersicon pennelli* to salt-dependant oxidative stress: the root antioxydative system. *Physiol Plant* 112: 487–494
15 Muscolo A, Sidari M, Panccio MR (2003) Tolerance of Kikuyu grass to long term salt stress in associated with induction of antioxidant defences. *Plant Growth Regul* 41: 57–62
16 Broetto F, Luttge U, Ratajczak R (2002) Influence of light intensity and salt-treatment on mode of photosynthesis and enzymes of the antioxidative response system of *Mesembryantheumum crystallinum*. *Funct Plant Biol* 29: 13–23
17 Parida AK, Das AB, Mohanty P (2004) Defence potentials to NaCl in a mangrove, Bruguiera parviflora: differential changes of isoforms of some antioxidative enzymes. *J Plant Physiol* 161: 531–542
18 Castillo F (1996) Antioxidative protection in the inducible CAM plant *Sedum album* L. following the imposition of severe water stress and recovery. *Oecol* 107: 469–477
19 Cheesman JM, Herendeen LB, Cheesman AT, Clough BF (1997) Photosynthesis and photorespiration in mangroves under field conditions. *Plant Cell Environ* 20: 579–588
20 Hernandez JA, Jimenez A, Mullineaux P, Sevilla F (2000) Tolerance of pea (*Pisum sativum* L.) to long-term salt stress is associated with induction of antioxidant defences. *Plant Cell Environ* 23: 853–862
21 Sreenivasula N, Grimm B, Wobus U, Wescke W (2000) Differential responses of antioxidant compounds to salinity stress in salt tolerant and salt sensitive seedlings of foxtail millet (*Setaria italica*). *Physiol Plant* 109: 435–441
22 Lee DH, Kim YS, Lee CB (2001) The inductive responses of the antioxidant enzymes by salt stress in the rice (*Oryza. Sativa* L.). *J Plant Physiol* 158: 737–745
23 Hurkman WJ, Fornia CS, Tanaka CK (1989) A comparison of the effects of salt on polypeptides and translatable mRNA in roots of a salt-tolerant and salt-sensitive cultivar of barley. *Plant Physiol* 90: 1444–1456

Biosaline Agriculture and Salinity Tolerance in Plants
Edited by M. Öztürk, Y. Waisel, M.A. Khan and G. Görk


# An overview of the coastal zone plant diversity and management strategies in the mediterranean region of Turkey

Münir Öztürk[1], Aykut Güvensen[2], Çiğdem Görk[3] and Güven Görk[3]

[1] *Ege University, Centre for Environmental Studies, 35100, Bornova-Izmir, Turkey*
[2] *Ege University, Biology Department, 35100, Bornova-Izmir, Turkey*
[3] *Mugla University, Biology Department, Mugla, Turkey*

## Introduction

The Mediterranean basin has served as the cradle for well known civilizations [1]. Its favourable environmental conditions have attracted humans towards the coastal parts for thousands of years and they exploited the land very severely, thus resulting in the degradation of this complex ecosystem [2]. During the last few decades the basin has become a focus of attention for scientists and several projects have been followed [3–6]. Turkey is one of the countries with a long coastline in the basin including the Mediterranean Sea and Aegean Sea with a total length of 5,191 km [4]. The high mountain ranges run in close proximity to the shoreline allowing some of the rivers to form fertile alluvial plains. Due to the varying geological features, the coastline is highly indented embodying several bays and coves serving as the main area for tourism and recreation as well as other coastal uses.

The area is endowed with a rich plant diversity. Out of these, the saline and dune vegetation occupy an important place due to their natural recreational values as well as importance in the production of food, fodder, fibre and many other products [7].

Much work has been done on the flora of Turkey [8], but very few papers have been published on the coastal zone plant cover [9, 11–13]. In this paper, an attempt is made to present an overview of the coastal zone plant cover of the saline and dune habitats in the Mediterranean phytogeographical region of Turkey and suggestions for the management of this fragile ecosystem.

## Study area

The study area covers the states of İzmir, Aydın, Denizli, Muğla, Antalya, İçel, Adana and Antakya located in the Mediterranean phytogeographgical region of Turkey extending from Canakkale up to the Syrian border and covering an area of nearly 500,000 km$^2$ with several varying biotopes [14].The area in general experiences a typical Mediterranean climate with dry-hot summers and mild-rainy winters (Fig. 1). The precipitation regimes of Izmir, Aydın, Denizli, Muğla, Antalya and İçel follow the course as WSASu (W: Winter; S: Spring; A: Autumn; Su: Summer). In Adana and Antakya it is observed as WASSu (Tab. 1) [15, 16].

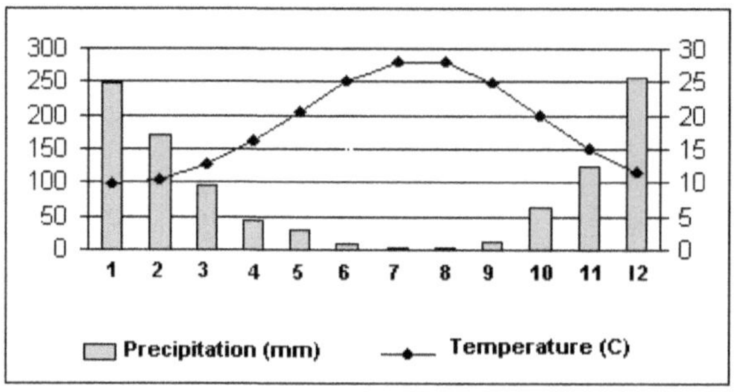


**Figure 1.** The temperature and precipitation values in the Mediterranean climate

**Table 1.** Precipitation regime of the study area (seasonal rainfall, mm) (1994–2000). From [16]

| States | Winter (W) | Spring (S) | Summer (Su) | Autumn (A) | Precipitation regimes |
|---|---|---|---|---|---|
| **İzmir** | 399.8 | 146.8 | 14.1 | 137.5 | WSASu |
| **Aydın** | 355.0 | 149.4 | 21.1 | 139.9 | WSASu |
| **Denizli** | 239.5 | 160.4 | 37.8 | 97.5 | WSASu |
| **Muğla** | 722.0 | 231.3 | 36.0 | 217.0 | WSASu |
| **Antalya** | 695.5 | 214.4 | 12.2 | 166.9 | WSASu |
| **İçel** | 356.7 | 127.5 | 25.8 | 114.4 | WSASu |
| **Adana** | 333.9 | 124.3 | 29.5 | 168.5 | WASSu |
| **Antakya** | 568.1 | 213.7 | 57.9 | 313.7 | WASSu |

## Plant diversity

The studies undertaken by several authors have revealed that different ecological habitats along the Mediterranean coastal zone embody an exceptionally high biodiversity [4, 9–13]. The pristine coastal dunes and beaches are of great importance as the breeding grounds of the endangered marine turtles *Caretta caretta* and *Chelonia mydas*. The area at the same time shows a high endemism in plants [4]. The plant taxa collected from the coastal zone of the Mediterranean area were identified by using the *Flora of Turkey and East Aegean Islands* [8] as well as other relevant books on the flora of Turkey. The records from other published papers were also noted [10–13].

**Table 2.** List of plant taxa from the Turkish Mediterranean coast

| Families | Taxa | Life form | Chorotype | Ecotype | Altitude (m above sl.) | Flowering time (months) |
|---|---|---|---|---|---|---|
| Amaryllidaceae | *Pancratium maritimum* | C | M | PM | 0–5 | 6–10 |
| Apiaceae | *Anethum graveolens* | T | IM | R | – | 6–7 |
| Apiaceae | *Bupleurum semicompotisum* | T | IM | H | 200 | 3–6 |
| Apiaceae | *B. euboeum* | T | M | H | 0–1500 | 6–8 |
| Apiaceae | **B. zoharii* | T | M | R | 0–1200 | 6–8 |
| Apiaceae | *Crithmum maritimum* | CH | IM | X | | 7–10 |
| Apiaceae | *Daucus broteri* | T | M | X | 0–100 | 6–8 |
| Apiaceae | *D. carota* | T | IM | X | 0–2000 | 6–9 |
| Apiaceae | *D. littoralis* | T | M | X | sea level | |
| Apiaceae | *Eryngium campestre* var. *virens* | H | IM | X | 0–1800 | 7–9 |
| Apiaceae | *E. maritimum* | H | IM | PM | sea level | 6–8 |
| Apiaceae | *Falcaria falcaroides* | H | IM | H | 380–1250 | 6–8 |
| Apiaceae | *Pseudorlaya pumila* | T | M | X | sea level | 3–6 |
| Asclepiadaceae | *Cionura erecta* | CH | IM | H | 0–1400 | 4–9 |
| Asclepiadaceae | *Cynanchum acutum* | C | IM | H | 0–1500 | 6–9 |
| Asteraceae | *Ambrosia maritima* | C | M | PM | 0–500 | 8–12 |
| Asteraceae | **Anthemis ammophila* | T | M | PM | 5 | 4–5 |
| Asteraceae | *Aster tripolium* | H | ES | PM | – | 6–9 |
| Asteraceae | *Atractylis carduus* | CH | IM | PM | sea level | 5 |
| Asteraceae | *Bellis annua* | T | M | H | 0–300 | 2–5 |
| Asteraceae | *B. perennis* | H | ES | H | 0–2000 | 3–8 |
| Asteraceae | *Carlina lanata* | T | M | X | 10–800 | 6–8 |
| Asteraceae | *Centaurea aegialophila* | H | M | X | 0–100 | 4–6 |
| Asteraceae | *C. calcitrapa* subsp. *calcitrapa* | H | M | R | 0–400 | 6–10 |
| Asteraceae | *C. spinosa* var. *spinosa* | CH | IM | PM | sea level | 6–7 |
| Asteraceae | *Chondrilla juncea* var. *juncea* | H | IM | X | 150–270 | 7–9 |
| Asteraceae | *Chrysanthemum coronarium* | T | M | R | 0–500 | 4–6 |
| Asteraceae | *Crepis foetida* subsp. *commutata* | T | IM | X | 0–1200 | 4–6 |
| Asteraceae | *C. foetida* subsp. *rhoeadifolia* | T | IM | PM | 0–2000 | 5–10 |
| Asteraceae | *C. sancta* | T | IM | H | – | – |
| Asteraceae | *Hedypnois cretica* | T | M | PM | 0–900 | 3–5 |
| Asteraceae | *Inula crithmoides* | CH | IM | H | sea level | 9–11 |
| Asteraceae | *I. graveolens* | T | M | PM | 0–800 | 8–10 |
| Asteraceae | *I. viscosa* | CH | M | X | 0–800 | 6–11 |
| Asteraceae | *Otanthus maritimus* | CH | M | PM | sea level | 5–11 |
| Asteraceae | *Xanthium strumarium* subsp. *cavanillesii* | T | IM | R | 0–950 | 7–10 |
| Boraginaceae | **Alkanna pinardii* | CH | M | X | 0–320 | 2–5 |

**Table 2.** Continued

| Families | Taxa | Life form | Chorotype | Ecotype | Altitude (m above sl.) | Flowering time (months) |
|---|---|---|---|---|---|---|
| Boraginaceae | *A. tinctoria* | CH | M | X | 0–800 | 4–7 |
| Boraginaceae | *Anchusa aggregata* | T | M | PM | 0–5 | 3–5 |
| Boraginaceae | *A. undulata* subsp. *hybrida* | C | M | PM | 0–900 | 5–7 |
| Boraginaceae | *Echium angustifolium* | CH | IM | PM | 0–870 | 8–12 |
| Boraginaceae | *Heliotropium europaeum* | T | M | R | 0–1400 | 6–9 |
| Brassicaceae | *Cakile maritima* | T | IM | PM | sea level | 6–8 |
| Brassicaceae | *Maresia nana* | T | IM | PM | sea level | 4 |
| Brassicaceae | *Raphanus raphanistrum* | T | IM | X | 0–400 | 3–5 |
| Brassicaceae | *Sinapis alba* | T | IM | R | 0–1400 | 2–4 |
| Brassicaceae | *S. arvensis* | T | IM | R | 0–1800 | 4–6 |
| Caryophyllaceae | **Arenaria pamphylica* var. *turcica* | T | IM | X | 0–800 | 4–8 |
| Caryophyllaceae | **Dianthus crinitus* var. *pamphylica* | T | IM | X | 0–215 | 3–5 |
| Caryophyllaceae | *Silene kotschyi* var. *maritima* | T | IM | PM | sea level | 5–7 |
| Caryophyllaceae | **S. pompeiopolitana* | T | M | PM | sea level | 4 |
| Caryophyllaceae | *Spergularia marina* | H | CSM | H | sea level | 3–6 |
| Caryophyllaceae | *S. media* | H | CSM | H | sea level | 5–8 |
| Caryophyllaceae | *S. rubra* | T | IM | H | 0–2500 | 4–8 |
| Chenopodiaceae | *Artrocnemum fruticosum* | CH | IM | H | 0–1100 | 8 |
| Chenopodiaceae | *A.glaucum* | CH | IM | H | | 10 |
| Chenopodiaceae | *Atriplex hastata* | T | IM | H | 0–50 | 5–8 |
| Chenopodiaceae | *A. lasiantha* | T | IM | R | 0–1800 | 5–7 |
| Chenopodiaceae | *Chenopodium album* subsp. *album* | T | IM | R | 0–2000 | 5–8 |
| Chenopodiaceae | *C. botrys* | T | IM | PM | 0–1900 | 5–7 |
| Chenopodiaceae | *C. murale* | T | IM | R | 0–400 | 5–7 |
| Chenopodiaceae | *C. opulifolium* | T | IM | R | sea level | 5–8 |
| Chenopodiaceae | *Halimione portulacoides* | CH | IM | H | 0–900 | 6–8 |
| Chenopodiaceae | *Halopeplis amplexicaulis* | T | IM | H | sea level | 6–8 |
| Chenopodiaceae | *Halocnemum strobilaceum* | CH | IM | H | 0–1100 | 7–9 |
| Chenopodiaceae | **Kalidiopsis wagenitzii* | CH | IM | H | | 5–6 |
| Chenopodiaceae | *Kochia prostrata* | CH | IM | X | 0–1900 | 6–8 |
| Chenopodiaceae | *Petrosimonia brachiata* | CH | IM | H | 0–900 | 6–9 |
| Chenopodiaceae | *Salicornia europaea* | T | IM | H | – | 7–9 |
| Chenopodiaceae | *Salsola kali* | T | IM | R | 0–1010 | 5–7 |
| Chenopodiaceae | *S. soda* | T | IM | H | sea level | 5–7 |
| Chenopodiaceae | *S. ruthenica* | CH | IM | PM | 0–1750 | 5–7 |
| Chenopodiaceae | *Suaeda prostrata* subsp. *prostrata* | T | IM | H | sea level | 6–9 |
| Cistaceae | *Fumana thymifolia* var. *thymifolia* | CH | IM | PM | 0–250 | 3–4 |

**Table 2.** Continued

| Families | Taxa | Life form | Chorotype | Ecotype | Altitude (m above sl.) | Flowering time (months) |
|---|---|---|---|---|---|---|
| Cistaceae | *Helianthemum stipulatum* | CH | M | PM | | 4 |
| Convolvulaceae | *Calystegia soldanella* | C | IM | PM | sea level | 5–7 |
| Convolvulaceae | *Convolvulus lanatus* | CH | Ss | PM | sea level | 5 |
| Convolvulaceae | *Cressa cretica* | T | IM | H | sea level | 6–8 |
| Convolvulaceae | *Ipomoea stolonifera* | T | M | PM | sea level | 6–9 |
| Cyperaceae | *Bolboschoenus maritimus* var. *maritimus* | C | CSM | H | 0–2000 | |
| Cyperaceae | *Carex divisa* | C | ES | H | 0–2800 | – |
| Cyperaceae | *Cyperus capitatus* | C | IM | H | 6–8 | 1–5 |
| Cyperaceae | *C. longus* | C | CSM | H | 0–1850 | 5–9 |
| Cyperaceae | *Eleocharis mitracarpa* | C | CSM | H | 0–2400 | 4–9 |
| Cyperaceae | *E. palustris* | C | CSM | H | 0–2400 | – |
| Cyperaceae | *Schoenoplectus litoralis* | C | IM | H | 0–1370 | 4–10 |
| Cyperaceae | *Schoenus nigricans* | H | IM | H | 0–2000 | 3–7 |
| Euphorbiaceae | *Euphorbia paralias* | H | M | PM | 0–10 | 4–9 |
| Euphorbiaceae | *E. peplis* | T | M | PM | sea level | 6–9 |
| Fabaceae | *Alhagi mannifera* | CH | IM | X | 0–2330 | 6–8 |
| Fabaceae | *A. pseudalhagi* | CH | IT | H | 0–1200 | 6–8 |
| Fabaceae | *Argyrolobium uniflorum* | H | Ss | PM | sea level | 5 |
| Fabaceae | *Astragalus epiglottis* | T | M | X | | 5 |
| Fabaceae | **A. suberosus* subsp. *mersinensis* | T | M | X | 0–1100 | 4–6 |
| Fabaceae | **A. suberosus* subsp. *suberosus* | T | M | X | 0–1500 | 4–6 |
| Fabaceae | *A. subuliferus* | CH | M | X | | 4 |
| Fabaceae | *Factorovskya aschersoniana* | T | M | X | sea level | 1–2 |
| Fabaceae | *Glycyrrhiza glabra* var. *glandulifera* | H | IM | PM | 0–1800 | 6–7 |
| Fabaceae | *Lotus corniculatus* var. *tenuifolius* | H | CSM | H | 0–2750 | 4–9 |
| Fabaceae | *Medicago littoralis* var. *littoralis* | T | IM | PM | sea level | 4–6 |
| Fabaceae | *M. marina* var. *marina* | H | IM | PM | sea level | 2–6 |
| Fabaceae | *M. minima* var. *minima* | T | IM | PM | 0–1750 | 3–5 |
| Fabaceae | *Melilotus indica* | T | IM | H | 0–1750 | 2–5 |
| Fabaceae | *M. messanensis* | T | M | PM | sea level | 2–4 |
| Fabaceae | *Ononis natrix* subsp. *hispanica* | CH | M | PM | sea level | 5–8 |
| Fabaceae | *O. variegata* | T | M | PM | sea level | 4–6 |
| Fabaceae | *Ornithopus compressus* | T | M | X | 0–300 | 4–6 |
| Fabaceae | *Prosopis farcta* | P | IM | H | 0–1450 | 6 |
| Fabaceae | *Trifolium campestre* | H | CSM | X | 0–2200 | 2–4(–9) |
| Fabaceae | *T. resupinatum* var. *resupinatum* | T | IM | H | 0–1500 | 5 |
| Fabaceae | *Trigonella cylindracea* | T | M | PM | sea level | 4–6 |

**Table 2.** Continued

| Families | Taxa | Life form | Chorotype | Ecotype | Altitude (m above sl.) | Flowering time (months) |
|---|---|---|---|---|---|---|
| Fabaceae | *T. halophila* | T | M | PM | sea level | 4–6 |
| Frankeniaceae | *Frankenia pulverulenta* | T | IM | H | 0–1000 | 7–8 |
| Gentianaceae | *Blackstonia perfoliata* subsp. *perfoliata* | T | IM | H | 0–900 | 4–10 |
| Gentianaceae | *Centaurium spicatum* | T | IM | H | 0–1070 | 7–8 |
| Gentianaceae | *C. erythraea* subsp. *erythraea* | H | ES | H | 0–900 | 5–8 |
| Gentianaceae | *C. tenuiflorum* subsp. *tenuiflorum* | T | IM | H | 0–1150 | 6–8 |
| Geraniaceae | *Geranium dissectum* | T | IM | X | 0–400 | 4–5 |
| Geraniaceae | *Erodium cicutarium* subsp. *bipinnatum* | T | IM | PM | sea level | 3–4 |
| Guttiferae | *Hypericum polyphyllum ssp. polyphyllum* | CH | M | X | 0–770 | 5–6 |
| Guttiferae | *H. triquetrifolium* | CH | IM | X | 0–1250 | 5–9 |
| Illecebraceae | *Paronychia argentea* var. *argentea* | H | IM | PM | sea level | 3–6 |
| Juncaceae | *Juncus acutus* | C | IM | H | 0–150 | 3–5 |
| Juncaceae | *J. maritimus* | C | IM | H | 0–1050 | 5–7 |
| Juncaceae | *J. subnodulosus* | C | IM | H | 0–100 | 6–7 |
| Juncaceae | *J. subulatus* | C | M | H | sea level | 4–6 |
| Lamiaceae | *Lavandula stoechas* subsp. *stoechas* | CH | M | X | sea level | 3–6 |
| Lamiaceae | *Prasium majus* | CH | M | X | 0–1300 | 3–7 |
| Lamiaceae | *Salvia viridis* | T | M | X | 0–1300 | 3–7 |
| Lamiaceae | *Teucrium polium* | H | CSM | PM | 0–2050 | 6–9 |
| Lamiaceae | *T. scordium* subsp. *scordioides* | C | ES | PM | 50–2350 | 5–9 |
| Liliaceae | *Asparagus aqutifolius* | CH | R | X | 0–1525 | 8–9 |
| Liliaceae | *Urginea maritima* | C | M | PM | 0–300 | 9–11 |
| Orchidaceae | *Orchis palustris* | C | CSM | H | 0–1950 | 6–7 |
| Papaveraceae | *Papaver stylatum* | T | M | X | 0–1200 | 3–6 |
| Plantaginaceae | *Plantago coronopus* subsp. *commutata* | T | M | PM | 0–800 | 2–7 |
| Plantaginaceae | *P. coronopus* subsp. *coronopus* | T | IM | H | 0–1750 | 4–11 |
| Plantaginaceae | *P. crassifolia* | H | M | PM | 0–900 | 5–10 |
| Plantaginaceae | *P. cretica* | T | M | X | 0–600 | 4–5 |
| Plantaginaceae | *P. lagopus* | H | M | H | 0–2000 | 4–8 |
| Plantaginaceae | *P. lanceolata* | H | IM | H | | |
| Plantaginaceae | *P. maritima* | H | IM | H | 0–2400 | 5–8 |
| Plantaginaceae | *P. scabra* | T | IM | X | 0–1250 | 5–11 |
| Plumbaginaceae | *Limonium angustifolium* | H | M | H | sea level | 5–10 |
| Plumbaginaceae | *L. bellidifolium* | H | ES | H | 0–1010 | 6–9 |
| Plumbaginaceae | *L. echioides* | T | M | H | 0–200 | 4–7 |
| Plumbaginaceae | **L. effesum* | H | M | H | 0–750 | 7–8 |
| Plumbaginaceae | *L. gmelinii* | H | ES | H | 0–1450 | 5–10 |

**Table 2.** Continued

| Families | Taxa | Life form | Chorotype | Ecotype | Altitude (m above sl.) | Flowering time (months) |
|---|---|---|---|---|---|---|
| Plumbaginaceae | *L. graecum* var. *graecum* | H | M | H | sea level | 5–7 |
| Plumbaginaceae | *L. ocymifolium* | H | M | H | sea level | 5–7 |
| Plumbaginaceae | *L. sieberi* | CH | M | H | 0–5 | 5–7 |
| Plumbaginaceae | *L. sinuatum* | H | M | H | 0–100 | 5–7(10) |
| Plumbaginaceae | *L. virgatum* | H | M | H | 0–20 | 6–10 |
| Poaceae | *Aeluropus littoralis* | C | IM | PM | 0–1200 | 5–10 |
| Poaceae | *Aira elegantissima* var. *elegantissima* | T | M | PM | 0–300 | 4–5 |
| Poaceae | **Alopecurus myosuroides* var. *latialatus* | T | M | H | sea level | 5 |
| Poaceae | *Ammophila arenaria* subsp. *arundinacea* | C | M | PM | sea level | 6–8 |
| Poaceae | *Briza maxima* | T | IM | X | 0–320 | 4–5 |
| Poaceae | *Bromus arvensis* | T | IM | H | 0–2900 | 6–8 |
| Poaceae | **B. psammophilus* | T | M | PM | sea level | 6 |
| Poaceae | *B. rubens* | T | IM | PM | 0–1000 | 3–6 |
| Poaceae | *B. tectorum* | T | CSM | PM | 0–2000 | 3–6 |
| Poaceae | *Catabrosa aquatica* | C | IM | H | 0–2600 | 5–8 |
| Poaceae | *Corynephorus divaricatus* | T | M | X | 0–1100 | 4–6 |
| Poaceae | *Crypsis aculeate* | T | IM | PM | 0–1510 | 6–10 |
| Poaceae | *Cutandia dichotoma* | T | IM | PM | sea level | 5 |
| Poaceae | *C. maritima* | T | M | PM | sea level | 5–6 |
| Poaceae | *Cynodon dactlylon* var. *dactylon* | C | IM | PM | 0–1830 | 4–9 |
| Poaceae | *Elymus factus* subsp. *farctus* var. *farctus* | C | M | PM | sea level | 6–8 |
| Poaceae | *Hordeum marinum* var. *marinum* | T | IM | PM | 0–830 | 5–6 |
| Poaceae | *H. marinum* var. *pubescens* | T | ES | H | 0–100 | 5–6 |
| Poaceae | *H. murinum* subsp. *glaucum* | T | IM | X | 0–1750 | 4–7 |
| Poaceae | *Imperata cylindrica* subsp. *cylindrica* | C | IM | PM | 0–760 | 4–7 |
| Poaceae | *Lagurus ovatus* | T | M | PM | 0–50 | 4–6 |
| Poaceae | *Lolium rigidum* var. *rigidum* | T | IM | H | 0–1850 | 4–7 |
| Poaceae | *Parapholis filiformis* | T | M | PM | sea level | 6 |
| Poaceae | *P. incurva* | T | IM | H | 0–100 | 4–7 |
| Poaceae | *P. canariensis* | T | IM | H | 0–1000 | 5–7 |
| Poaceae | *Phleum exaratum* subsp. *exaratum* | T | IM | H | 0–2300 | 5–7 |
| Poaceae | *Phragmites australis* | C | ES | H | 0–2400 | 8–10 |
| Poaceae | *Poa bulbosa* | C | IM | X | 0–3000 | 5–7 |
| Poaceae | *P. trivialis* | C | IM | H | 0–2210 | 5–8 |

**Table 2.** Continued

| Families | Taxa | Life form | Chorotype | Ecotype | Altitude (m above sl.) | Flowering time (months) |
|---|---|---|---|---|---|---|
| Poaceae | *Polypogon maritimus* subsp. *maritimus* | T | ES | PM | 0–400 | 5–6 |
| Poaceae | *P. monspeliensis* | T | ES | H | 0–1200 | 4–8 |
| Poaceae | *Prapholis incurva* | T | IM | H | 0–100 | 4–7 |
| Poaceae | *Puccinellia festuciformis* | H | IM | H | sea level | 5–6 |
| Poaceae | *Sporobolus virginicus* | C | IM | H | 0–950 | 4–10 |
| Poaceae | *Triplachne nitens* | T | M | PM | sea level | 5 |
| Poaceae | *Vulpia fasciculata* | T | M | PM | 0–20 | 5–6 |
| Polygonaceae | *Polygonum aviculare* | T | CSM | PM | 0–700 | 7–11 |
| Polygonaceae | *P. equisetiforme* | H | IM | X | sea level | 6–9 |
| Polygonaceae | *P. maritimum* | H | IM | PM | sea level | 6–11 |
| Polygonaceae | *Rumex bucephalophorus* | T | M | X | 0–150 | 2–5 |
| Primulaceae | *Anagallis arvensis* var. *arvensis* | T | IM | PM | 0–1400 | 4–9 |
| Primulaceae | *Glaux maritima* | C | IM | H | 0–1720 | 5–8 |
| Primulaceae | *Lysimachia atropurpurea* | T | M | R | 0–1000 | 5–10 |
| Primulaceae | *Samolus valerandi* | H | CSM | H | 0–900 | 5–9 |
| Ranunculaceae | *Ranunculus marginatus* var. *marginatus* | T | CSM | H | 0–850 | 3–6 |
| Rosaceae | *Rubus sanctus* | CH | IM | X | 0–1250 | 6–8 |
| Rubiaceae | *Oldenlandia capensis* | T | IM | H | sea level | 9 |
| Scrophulariaceae | *Rhamphicarpa medwedewii* | T | IM | H | sea level | 8 |
| Scrophulariaceae | *Scrophularia canina* | CH | M | H | 0–1500 | 4–7 |
| Scrophulariaceae | *Verbascum sinuatum* var. *adenosepalum* | H | M | PM | 0–1100 | 5–10 |
| Scrophulariaceae | *V. sinuatum* var. *sinuatum* | H | M | PM | 0–1100 | 5–10 |
| Solanaceae | *Solanum alatum* | H | IM | X | 0–1350 | 6–11 |
| Tamaricaceae | *Tamarix hampeana* | P | IM | H | – | 4 |
| Tamaricaceae | *T. parviflora* | P | M | H | 0–300 | 3–6 |
| Tamaricaceae | *T. smyrnensis* | P | IM | H | 0–1000 | 4–8 |
| Tamaricaceae | *T. tetrandra* | P | IM | H | 0–1300 | 5 |
| Thymelaeaceae | *Thymelaea hirsuta* | CH | M | X | sea level | 9–12 |
| Verbenaceae | *Phyla nodiflora* | CH | IM | PM | sea level | 4–8 |
| Verbenaceae | *Phyla nodiflora* | CH | IM | PM | sea level | 4–8 |
| Verbenaceae | *Verbena officinalis* | CH | IM | X | 0–1800 | 6–8 |
| Zygophyllaceae | *Tribulus terrestris* | T | IM | X | 0–1200 | 6–9 |

**IT:** Irano-Turanian, **ES:** Euro-Siberian, **M:** Mediterranean, **IM:** Imperfectly known, **Ss:** Sahoro-Sindian, **CSM:** Common, **ID:** Indifferent, **H:** Hemicryptophytes, **CH:** Chamaephytes,**T:** Therophytes, **C:** Cryptophytes,**P:** Phanerophytes, **PM:** Psammophytes, **H:** Halophytes, **X:** Xerophytes, **R:** Ruderals, *: Endemics

During the present survey 213 taxa belonging to 37 families were recorded to show a distribution along the Turkish Mediterranean coast (Tab. 2). The four major families embodying the highest number of taxa are Poaceae (26 taxa), Asteraceae (21 taxa), Fabaceae (23 taxa) and Chenopodiaceae (19 taxa) (Tab. 2). A similar situation is seen in the neighbouring Mediterranean country of Greece [17]. The genera *Limonium, Plantago, Tamarix, Chenopodium, Juncus, Bromus* and *Astragalus* have the highest number of taxa (Tab. 2). The distribution of different ecological forms is as follows; halophytes (85 taxa), psammophytes (67 taxa), xerophytes (47 taxa) and ruderals (14 taxa) (Tab. 2). Phytogeographically, 51.17 % of these taxa are imperfectly known, 35.69 % Mediterranean, 6.57 % Cosmopolite, 5.16 % Euro-Siberian, 0.94 % Sahoro-Sindian and 0.47 % Irano-Turanian. The life form spectrum reveals that 103 taxa of the coastal plant cover are therophytes, 39 taxa are hemicryptophytes, 36 taxa are chamaephytes, 30 taxa are cryptophytes and 5 taxa are phanerophytes (Tab. 2). Out of the 213 taxa, 13 are endemics. The red data book of Turkish plants reveals that 13 endemics and 10 non-endemic taxa from the coastal plant cover are in danger of becoming extinct [18].

## Soils and groundwater

The coastal zones abound in sand dunes and saline habitats. Soils and groundwater samples collected from the coastal zone cities were subjected to a detailed physical and chemical analysis using the methods outlined in detail in [19, 20].

In summers, soils are highly saline with E.C values ranging between 30.53 and 67.95 dS/m. The pH varies between 7.46 and 9.09, being strongly alkaline to slightly alkaline, the dominant cation is sodium (367.1–714.32 me/lt), the dominant anion chloride (338.23–871.67 me/lt) and boron values lie between 0.57 and 2.729 ppm. E.C values in the groundwaters during summer lie between 31.17 and 79.98 dS/m, being highly saline, pH is 7.01–7.51, being neutral to slightly alkaline. The dominant cation is sodium (281.76–795.7 me/lt), dominant anion is chloride (329.7–952.52 me/lt) and the boron content is very high (1.4–3.52 ppm).

E.C values in the soils during winter vary between 3.95 and 36.95 dS/m, being saline, medium-saline and highly saline, pH being 7.67–9.10 and these are strongly alkaline, medium alkaline to slightly alkaline; the dominant cation is sodium (38.7–415.3 me/lt), the dominant anion is chloride (25.0–451.5 me/lt) and the boron values are higher (1.47–4.5 ppm). In the groundwaters, E.C values in winter lie between 7.36 and 85.08 dS/m, being slightly saline to highly saline. The pH varies between 7.02 and 8.28 and these are neutral, slightly alkaline, medium alkaline; the dominant cation is sodium (70.2–976. me/lt), the dominant anion is chloride (61.2–1083.15 me/lt) and the boron content is very high (0.98–4.75 ppm).

E.C values in the soils during spring are in the range of 8.57–49.48 dS/m, being slightly saline to highly saline; the pH lies between 7.72–9.03, being strongly alkaline, slightly alkaline to medium alkaline; the dominant cation is sodium (79.15–593.02 me/lt), the dominant anion is chloride (67.45–635.07 me/lt) and the boron content is high (1.9–5.3 ppm). The E.C values in the groundwaters in spring lie between 17.85 and 93.59 dS/m, these are medium saline to highly saline; the pH lies between 7.11 and 7.92, being slightly alkaline to medium alkaline; the dominant cation is sodium (102.6–1147.2 me/lt), the dominant anion is chloride (107.2–1223.12 me/lt) and the boron content ranges between 0.88 and 5.02 ppm.

E.C values in the groundwaters during autumn vary between 31.04 and 69.74 dS/m and all samples are highly saline. The pH of the samples lies between 7.49 and 8.25, being medium alkaline to slightly alkaline. The dominant cation is sodium (331.5–745.45 me/lt), dominant anion is chloride (295.8–1004.25 me/lt) and the boron values lie between 0.9–3.85 ppm.

Some investigations on the salinity–alkalinity problems of the Turkish Mediterranean coastal zone cities have been carried out earlier [21]. The saline and sodic soils in the Mediterranean region cover an area of 560,000 ha. In Muğla, Antakya and Antalya states, these vary between 5,000 and 50,000 ha; in Izmir, Aydın and İçel between 50,000 and 100,000 ha, and in Adana over 100,000 ha. The saline areas are greater in size than its neighbouring Mediterranean country Syria (532,000 ha).

## Land degradation activities

The biotic pressures significantly damaging the coastal zones are demographic developments and urbanization, summer houses, exploitation for tourism and recreation, salinity–alkalinity problems, industrial activities, reclamation of land from dunes and wetlands and sand extraction [4, 22]. A large number of marshes on the coast have been changed into touristic resorts or agricultural areas. The climate, soils and vegetation interact strongly with such activities. There has been a steady migration during the last few decades towards the socio-economically developed coastal zones such as the Mediterranean and Aegean regions, leading to overpopulation and heavy urbanization in the area [4]. Both these regions have experienced a high tourism activity. The attractive coastal cities from Kuşadası to Alanya hosted 11.6 million foreign tourists in 2004. This, together with the illegal invasion of the coastal zone for cultivation, heavy pressure of pollutants brought from inland areas, untreated wastewater resulting from urbanization and industry and summer houses has resulted in the degradation reaching a stage where negative impact is currently limited but imminently significant [23, 24]. These areas use high amounts of freshwater, pesticides and fertilisers resulting in the deterioration of this ecosystem.

## Management strategies

Throughout history, the coastal zones were exploited and disputed by their inhabitants. The situation in the Mediterranean phytogeographical region of Turkey is no different. But fortunately today people have started to realize that it is necessary to stop abusing this fragile environment. However, for the management of these areas, a broad range of information spanning different fields is needed. The dynamic management response will be possible by a close monitoring of the status of coastal ecosystems, quality of the habitats and economic indicators [23–25]. For this purpose, sound information on plant diversity, vegetation cover, habitat types, and locations of species-communities should be collected and land cover maps should be prepared to note the status of habitat deterioration caused by climate fluctuations and by human impact [26–28]. The plant diversity representing various life forms together with species richness and other biodiversity indicators will help in the determination of site quality as presented in this work. Salinization and alkalization problems in the coastal states are also increasing [21]. The percentage of soils with high carbonate concentrations is very high and soils with a pH value in excess of 8 are 4.7 %. An ecologically sound way to cope with this situation would be the use of halophytic taxa in such areas. The halophytes could be evaluated as potential agricultural crops by growing them on saline soils, in particular along the coastal parts where seawater is available for irrigation, instead of destroying them as a wasteful group of plants. However, this will rely on a high degree of salt tolerance, not only of the perennial species used to lower a saline water table, but also of the crops to follow, as some salt will remain in the soil. The species like *Salicornia europaea* and *Suaeda maritima* may be a good choice for such areas. In fact there is a great potential existing in the halophytic plant cover for consumption as well as amelioration of the degraded lands along the coastal zones [10, 26]. The coastal zone plant cover is mainly composed of psammophytes and halophytes. They can be used for biological desalination and reclamation of saline–alkaline habitats [7]. Remote sensing and GIS (Geographic Information Systems) should be used in the mapping and monitoring of coastal ecosystem [24, 28]. A multidisciplinary program following a science based approach for methods, standards, data collection, research networks and development of land use models incorporating natural/human induced factors should be started for the evaluation of ecologically sustainable economic productivity potential of the coastal zone areas. The overlapping of responsibilities among different authorities in the region should immediately be solved. These measures can be used for designing conservation policies. For this purpose both national decision-makers and non-governmental organizations, as well as communities living in the area, should join hands to overcome this problem.

## References

1 Dalman PR (1998) *Plant life in the world's Mediterranean climates*. University of California Press, Berkeley, California 258

2 Belfiore S (1996) The role of the European community in the Mediterranean coastal zone management. *Ocean Coast Manage* 31: 219–258

3 Vertovec M, Sakcali S, Ozturk M, Salleo S, Giacomich P, Feoli E, Nardini A (2001) Diagonosing plant water status as a tool for quantifying water stress on a regional basis in Mediterranean drylands. *Ann For Sci* 58: 113–125

4 Öztürk M, Çelik A, Yarci C, Aksoy A, Feoli E (2002) An overview of plant diversity, land use and degradation in the Mediterranean region of Turkey. *Environ Manage Health* 13: 442–449

5 Feoli E, Giacomich P, Mignozzi K, Ozturk M, Scimone M (2003) Monitoring desertification risk with a index integrating climatic and remotely-sensed data: An example from the coastal area of Turkey. *Management of Environmental Quality: An International Journal* 14: 10–21

6 Berberoğlu S (2003) Sustainable management for the Eastern Mediterranean Coast of Turkey. *Environ Manage* 31: 442–451

7 Öztürk M, Güvensen A (2002) Studies on the halophytic vegetation of saline-alkaline habitats in West Anatolia-Turkey. In: AM Hegazi et al. (eds): *Proc. of the International Synposium on Optimum Resources Utilization in Salt-Effects Ecosystems in Arid and Semiarid Regions* DRC, Cairo, Egypt 241–244

8 Davis PH (1965–1988) *Flora of Turkey and East Aegean Island.* Edinburgh University Press, UK 1–10
9 Zeybek N, Sheikh KH, Ozturk MA (1976) Coastal halophytes of the Aegean region. Cento Symposium, Adana 148–153
10 Gehu JM, Uslu T, Costa M (1989) Apport a la Connaissance Phytosociologique du Littoral Sud de la Turquie Mediterraneenne. *Colloques phytosociologiques* XIX: 591–622
11 Lovric AZ, Uslu T (1986) Dry coastal ecosysytems of Turkey. *Ecosystems of the World* II-A, 4/12. Elsevier, Amsterdam 443–469
12 Guvensen A, Ozturk M (2003) Halophytic plant diversity of south Aegean coastal zone in Turkey. *Pak J Bot* 35: 853–864
13 Çakan H, Yilmaz KT, Düzenli A (2005) First comprehensive assessment of the conservation status of the flora of the Çukurova Deltas, southern Turkey. *Oryx* 39/1: 1–4
14 Berberoğlu S, Yilmaz KT, Özkan C (2004) Mapping and monitoring of coastal wetlands of Cukurova Delta in the East Mediterranean Region. *Biodivers Conserv* 13: 615–633
15 Akman Y, Daget PH (1971) Quelques aspects synoptiques des climats de la Turquie. *Bull Soc Lang Geogr* 5: 269–300
16 Akman Y (1999) İklim ve biyoiklim (Biyoiklim metodları ve Türkiye iklimleri). Kariyer Press, Istanbul
17 Babalonas D, Syroka KV, Papastergiadou ES (1995) Review of plant communities from Greek dunes and salt-marshes, a preminary summarizing list. *Annali Di Botanica* 3: 107–117
18 Ekim T, Koyuncu M, Vural M, Duman H, Aytac Z, Adıguzel N (2000) *Red data book of Turkish plants.* Bariscan Press, Ankara 246
19 Richards LA (1954) Diagnosis and improvement of saline and alkali soils. United States Laboratory Staff, Agriculture Handbook No: 60, Soil and Water Conservation Research Branch Agricultural Research Service, Riverside, California, USA 69–180
20 Soil Survey Staff (1951) *Soil survey manual, U.S.D.A Handbook*, Washington DC Press, USA 18
21 Öztürk M, Akbaş F, Güvensen A, Fırattekin Y (2000) Studies on the edaphic relations in Söke. 'Plant Life in Sout-West and Central Asia, Vth International Symposium', Tashkent-Uzbekistan 139–148
22 Conacher AC, Sala M (1998) *Land Degradation in the Mediterranean Environments of the World: Nature and Extent, Causes and Solutions.* Wiley, Chichester, UK
23 Yilmaz KT (1998) Ecological diversity of the Eastern Mediterranean region of Turkey and its conservation. *Biodivers Conserv* 7: 87–96
24 Berberoğlu S, Alphan H, Yilmaz KT (2003) A remote sensing approach for detecting agricultural encroachment on the Eastern Mediterranean Coastal Dunes of Turkey. *Turk J Agric For* 27: 135–144
25 Cori BS (1999) Spatial dynamics of Mediterranean coastal region. *J Coastal Conserv* 5: 105–112
26 Yilmaz KT (2002) Evaluation of the phytosociological data as a tool for indicating coastal dune degradation. *Israel J Plant Sci* 50: 229–238
27 Hoffman J (1998) Assessing the effects of environmental changes in a landscape by means of ecological characteristcs of plant species. *Landcape and Urban Planning* 41: 239–248
28 Donoghue DNM, Mironnet N (2002) Development of an integrated geographical information system prototype for coastal habitat monitoring. *Computers and Geosciencesa* 28: 129–141

Biosaline Agriculture and Salinity Tolerance in Plants
Edited by M. Öztürk, Y. Waisel, M.A. Khan and G. Görk


# Saline and alkaline vegetation of NE Africa and the Arabian peninsula: An overview

Shahina A. Ghazanfar

*Royal Botanic Gardens, Kew, Richmond, Surrey, TW9 3AB, UK*

## Introduction

The saline and alkaline vegetation of NE Africa and Arabia is influenced by the diverse geomorphology of the coastline and inland arid plains, and by the distribution patterns of tropical and extra-tropical plant species. A zonation of species from the sea or lake landwards is distinct and is determined by the declining influence of salinity and/or alkalinity, duration and degree of inundation, and structure of the substrate.

One of the most prominent features of saline or alkaline vegetation is that it is impoverished in the number of species and hardly exceeds 300 taxa. The main plant families, to which these belong, include Acanthaceae, Amaranthaceae, Asteraceae, Chenopodiaceae, Fabaceae, Plumbaginaceae, Polygonaceae and Zygophyllaceae. As the area is arid with high daytime temperatures, the majority of the species present show C4 photorespiration pathway thus allowing them to survive in these high temperatures, intense light and low moisture conditions.

Another feature of saline or alkaline vegetation is that much of it is composed of monospecific stands though mosaics with two or three species are also found. The majority of the species are obligate halophytes, but salt tolerant and salt secreting species such as *Sporobolus*, *Aleuropus*, *Limonium* and mangroves are also present. In tropical East Africa the arid and saline species are predominantly annuals while in Arabia they are chiefly succulents and hemicryptophytes.

In this paper I review the halophytic and glycophytic vegetation of NE Africa and Arabia, describe the main plant communities and the typical vegetation zonation patterns, significant differences in the tropical and extra-tropical flora and vicariance in the Arabian halophytic flora.

## Geographical area

The study area consists of the north eastern part of Kenya, NE and coastal south-east Yemen, coastal and inland Oman, the Saudi Arabian desert region, and coastal regions of the UAE, Qatar and Kuwait.

In tropical East Africa halophytic and glycophytic vegetation is limited mostly to lake basins in the Eastern Rift (mainly in the Kenyan Rift Valley) and a few coastal areas. A relatively large inland salt lake lies in north Kenya, west of Lake Turkana, associated with the Chalbi desert (a former lake).

The 8,000 km long coastline of the Arabian Peninsula shows a diverse topography of low sandy coastal dunes, flat, silty-saline depressions, and cliffs and littoral mountains. Climate throughout the Arabian Peninsula is arid with very little rainfall [46], and therefore there are large inland areas of high salinity with obligate halophytic vegetation.

The present review is based on publications by various authorities on tropical East Africa and the Arabian Peninsula, and my own research in the Sultanate of Oman. Various aspects of the vegetation of tropical East Africa has been described by Bogdan [1], Vesey-FitzGerald [2], Gillett et al. [3], Knapp [4], Clayton et al. [5], Lind and Morrison [6], and White [7]. A comprehensive account of the vegetation of the Arabian Peninsula is given in Ghazanfar and Fisher [8] with details of the coastal and halophytic vegetation by Deil [9 and references therein], Mandaville [10 and references therein], Ghazanfar [11, 12 and references therein].

## Halophytic vegetation of tropical East Africa

The desert vegetation of tropical East Africa (northern Kenya) falls under the Somalia-Masai regional centre of endemism which includes NE Uganda (Karamoja), most of Kenya between the Highlands and the coastal belt, and the dry lowlands of north and central Tanzania. The vegetation is characterised by an *Acacia–Commiphora* deciduous bushland and thicket, and grassland. Two genera, *Drake-Brockmania* and *Dasysphaera* are endemic to this floristic region.

The dry and halophytic vegetation is found primarily in the Chalbi desert, which is sparsely vegetated at the edges where there is subsurface water flow from springs from the surrounding mountains. Due to extensive evaporation the entire area is hyper-arid, and after seasonal rains and flood, the soil becomes highly saline supporting a few halophytic grasses which grow near runnels and water outlets.

In the surrounding plains and parts of the Chalbi desert the most extensive vegetation type is the semi-desert annual grassland. The main halophytic grass is *Drake-Brockmania somalensis*, a mat-forming annual, spreading by stolons. It occupies seasonally flooded places in silty and saline soils. *Drake-Brockmania somalensis* is distributed from Tanzania through to NE Africa (Sudan, Somalia, Ethiopia; excluding Uganda); it is also found on the Farasan Island (Saudi Arabia) in the Red Sea [13].

By far the most common and widespread in the desert and semi-desert are the annual grasses, *Aristida adscensionis* and *A. mutabilis*. These occupy the driest areas, extending up to elevations of 1,000 m on the drier parts of the surrounding hills. They occur on poor, shallow soils and during periods of drought may be absent for as long as the drought lasts [7]. Associated with these are a few species of subshrubs which form the perennial woody components of the sabkhat. Dominant among these are *Duosperma eremophilum* (Family Acanthaceae) and *Indigofera spinosa* (Family Leguminosae); the former found on relatively moister soils, and the latter on the drier soils.

Among other shrubs, *Lagenantha nogalensis* (Family Chenopodiaceae) a succulent that is tolerant of gypsophilous soils, forms almost pure stands on white calcareous soils of the old Chalbi lake bed; *Volkensinia prostrata* (Family Amaranthaceae) occurs on saline or alkaline soils at the margins of Lake Turkana and at the edges of the Chalbi desert [7].

Only a few trees or large shrubs are associated with arid and saline soils. These are usually stunted, with *Acacia reficiens* as the dominant species.

The main species around the saline lakes in Kenya and Uganda are *Cyperus laevigatus, Sporobolus spicatus* and *Dactyloctenium* spp. The vegetation of the grassland on the flats close to Kiboko river is described by Bogdan [1], which shows different dominant species inhabiting soils with different concentrations of salts: thus in slightly alkaline conditions the grass cover is mainly *Cenchrus ciliaris,* but as alkalinity increases *Chloris gayana* becomes dominant. With increasing moisture and alkalinity *Sporobolus robusta* becomes the dominant species and where soils are shallow, *Cynodon dactylon* appears, and in highly alkaline soils that are waterlogged in the rainy season, *Sporobolus spicatus* takes over and forms dense growth. In the flat valleys in the drier parts of Tanzania, such as the Pangani river valley, where the water is saline, the flood plains are dominated by grasses *Sporobolus robusta*, and shrubs such as *Suaeda monoica, Sesbania sesban, Salvadora persica* and *Triplocephalum holstii*.

The vegetation of the Lake Rukwa basin in Tanzania, which is chiefly grassland has been described by Vesey-FitzGerald [2]; this is summarised here: the vegetation can be divided into three zones from the fringe of the lake to the lake bed itself. The edges of the lake are occupied by almost pure stands of *Sporobolus robusta.* This species does not grow in water, but colonises the alkaline soils on the beach of the lake. The shallowly flooded alkaline swamp present on the extensive flat lake bed is vegetated by a single species *Diplachne fusca.* This is a rhizomatous perennial rooting and branching from the lower nodes forming dense mats over large areas. On the lake bed, the alkaline flats are occupied by two species: when the lake is dry *Spobolus spicatus* grows associated with *Psilolemma jaegeri*; when the flats are flooded with the highly alkaline water, *Psilolemma jaegeri* takes over and replaces *Sporobolus*.

Other saline and alkaline patches of vegetation exists around Lake Amboseli in Kenya, where the major vegetation type is the *Commiphora–Acacia* shrubland, and the saline and alkaline plains are dominated by *Suaeda monoica* and *Salvadora persica*. Alkaline grassland dominated by *Sporobolus spicatus* is also present around Lake Manyara. On the coast *Suaeda monoica* may form monospeciefic stands at the edges of the barren sand flats, while *Arthrocnemum indicum, Sporobolus virginicus* and *Suaeda monoica* grow on slightly elevated soil. *Salicornia pachystachya* may form carpets in summer. Other species include *Sesuvium portulacastrum* in the more moist areas.

The dry *Acacia–Commiphora* shrubland continues through from NE Africa to SW Arabia, and in Somalia the associated species on the the coastal plain include *Aerva javanica, Jatropha pelargoniifolia, Farsetia longisiliqua.* On gypsophilous soils succulent euphorbias (*E. coluumnaris, E. sepulta, E. mosaica*), *Dorstenia gypsophila, Pelargonium cristophoranum* are present [14].

## Halophytic vegetaion of the Arabian Peninsula

The main species in saline habitats are mostly perennial succulents, subshrubs and stoloniferous hemicryptophytes. Among the annuals, succulents species such as *Bienertia cycloptera* and *Zygophyllum simplex* are rare. The most common coastal and salt tolerant species are *Arthrocnemum macrostachyum*, *Halocnemum strobilaceum*, *Halopeplis perfoliata*, *Limonium* spp, *Salsola* spp, *Salicornia europaea*, *Seidlitzia rosmarinus*, *Suaeda* spp, *Zygophyllum* spp; the grasses and sedges *Aeluropus lagopoides*, *Juncus rigidus*, *Odyssea mucronata*, *Sporobolus spicatus*, *S. consimilis*, and mangroves *Avicennia marina*.

The typical plant communities of the coastal vegetation have been described from the Gulf of Aqaba [15–17], the Red Sea coast north of Jeddah [18], Tihama Coast [19], the Gulf of Aden [20–22], the Hadhramaut coast at Felek, east of Mukalla [22], the coastal, inland sabkha, and saline and brackish water vegetation of Oman [12, 23–26], the coastal vegetation of the islands of Masirah and Shagaf 27], vegetation of the Qurm Nature Reserve near Muscat [28, 29], coastal vegetation near Dubai [30], halophytic vegetation of Qatar [31–37], Bahrain [38–40], and Kuwait [41–43].

A summary of the typical vegetation of the Arabian Peninsula given here is taken from Deil [9]. For a detailed study see Deil [9 and references therein].

For the coastal vegetation, in the Gulf of Aqaba, generally the first zone consists of *Avicennia marina* on mud deposits followed by a *Limonium axillare* zone and then by a *Nitraria retusa–Zygophyllum album* zone. *Sueda monoica–S. vermiculata* zone can be seen near the shoreline in some locations. Where the coast is frequently inundated by the sea resulting in the top soil to be high in salt and the water table shallow (30–70 cm), *Arthrocnemum glaucum* is present. Where there is no inundation during high tide, a sterile zone is normally present followed by a *Suaeda pruinosa* zone. On the eastern coast of the Gulf of Aqaba a *Nitraria retusa* zone is present where the water table is at 100–140 cm, with the associated species *Zygophyllum album*, *Z. coccineum* and *Tamarix* spp. The last zone, a *Zygophyllum coccineum* zone, occurs on coarse textured sand with *Cyperus conglomeratus* and *Fagonia bruguieri* as associates. *Hyphaene thebaica* has its northernmost distribution near Eilat. The southernmost distributional limit of *Limonium pruinosum* is the southern part of the Gulf of Aqaba, replaced further south by *L. axillare*. A *Salvadora persica* open shrubland occurs on sand mounds and in alluvial fans where fresh water is close to the surface, and an open shrubland with *Nitraria retusa* and *Zygophyllum album* is present on aeolian sands where the groundwater is salty.

Along the Yemeni Red Sea coast northwards from Wadi Siham *Avicennia marina* zone is followed by a *Limonium cylindrifolium–Suaeda fruticosa–Limonium axillare* community which forms hummocks. A sterile sabkha is present, after which raised beaches above the high tide level are covered by *Atriplex farinosa*, *Zygophyllum hamiense*, *Aeluropus lagopoides* and *Halopyrum mucronatum*. Sand dunes towards the seaward side are colonised by *Suaeda monoica* and *Salsola spinescens*, and the inland dunes by *Odyssea mucronata*, *Jatropha pelargoniifolia* and *Leptadenia pyrotechnica*.

The southwestern corner of the Arabian Peninsula is characterised by the occurrence of a new coastal species, *Odyssea mucronata*, endemic to this part of Arabia. *O.*

*mucronata* is a clump-forming, spiny, rhizomatous perennial which colonises semi-mobile dunes and flat sandy areas. Depending on the depth of sand, an *Odyssea mucronata–Suaeda monoica* community can be distinguished on flat sandy layers overlying saline silts, and an *Odyssea mucronata–Panicum turgidum* community on deeper sand.

The Hadhramaut coast is situated in the transition zone from the southeastern to the southwestern vegetation type. This is seen from the *Cyperus conglomeratus* associations, where the Omano-Makranian element [44], *Coelachyrum piercei* and the Eritreo-Arabian element *Odyssea mucronata* are common members. The coastal vegetation shows a strong phytogeographical relationship with the coasts of northeast Africa. The species zones are: 1) coastal dunes colonised by sedges and grasses (*Cyperus conglomeratus, Halopyrum mucronatum, Odyssea mucronata, Coelachyrum piercei* and *Panicum turgidum*); 2) sandy-salty depressions colonised by the endemic *Urochondra setulosa* association, with the co-dominant *Arthrophytum macrostachyum, Limonium cylindrifolium* and *Crotalaria saltiana*; 3) clayey-salty, relatively wet areas colonised by monospecific stands of *Arthrophytum macrostachyum*; 4) sandy coastal plains colonised by the endemic *Anabasis ehrenbergii–Pulicaria hadramautica–Zygophyllum hamiense* association; 5) the karstic limestone plateau colonised by *Stipagrostis paradisea, Commiphora gileadensis* and *Euphorbia rubriseminalis*.

In the coastal vegetation of Oman four plant communities can be recognised: 1) A *Limonium stocksii–Zygophyllum qatarense* community in northern Oman where the coasts are mainly sandy and interspersed with rocky limestone headlands; 2) a *Limonium sarcophyllum–Suaeda aegyptiaca* community characteristic of rocky shores with narrow beach areas and a wide spray zone; 3) an *Atriplex–Suaeda* community characteristic of the vegetation of offshore islands, flat sandy beaches and coastal sabkhas (dominant and associated species are *Atriplex coriacea, A. farinosa, A. leucoclada, Arthrocnemum macrostachyum, Suaeda aegyptiaca, S. vermiculata, S. monoica, S. moschata* and *Halocnemum strobilaceum*), and a *Limonium axillare–Sporobolus–Urochondra* community characteristic of the vegetation of the southern coasts, with *Limonium axillare, Urochondra setulosa* and *Sporobolus* spp associated with several other species depending on coastal geomorphology; 4) coastal lagoons with *Sporobolus virginicus, S. iocladus* and *Paspalum vaginatum* as the main bordering species, and *Phragmites australis* and *Typha* spp forming the bordering reeds. In addition, *Avicennia marina* occurs throughout coastal Oman in discontinuous patches and over a wide range of water salinities [24].

On the Barr al Hikman Peninsula and the offshore island of Masirah, *Avicennia marina* is present in sheltered lagoons, a halophytic shrub community dominated by *Atriplex farinosa* and *Suaeda moschata* occurs on low coastal dunes which receive salt spray, and a *Halopyrum mucronatum–Urochondra setulosa* community occurs on more or less stabilised dunes. An *Arthrocnemum macrostachyum–Suaeda vermiculata* community occurs on the saline, silt plains and a *Limonium stocksii–Cyperus conglomeratus–Sphaerocoma aucheri* community on shallow sands.

A transect through the coastal dunes and sabkha in the UAE shows the typical dry haloseries within the Omano-Makranian region of the Arabian Gulf; four plant communities are present associated in the *Limonium stocksii–Zygophyllum qatarense* vegetation complex: 1) the seaward dunes colonised by the *Cornulaca monacantha–Sphaerocoma aucheri* community (the Salsolo-Suaedetalia of Knapp [45]); 2) the landward dunes colonised by *Halopyrum mucronatum* (stabilising the sand), *Atriplex leucoclada* and *Suaeda aegyptiaca*; 3) salty depressions which may be temporarily inundated with seawater colonised by *Halopeplis perfoliata*; 4) an ephemeral, salt tolerant *Frankenia pulverulenta–Zygophyllum simplex* plant community growing in depressions with sandy overlays. The landward dunes, away from the influence of salt spray, are also dominated by *Cornulaca monacantha* and *Sphaerocoma aucheri*. They are associated here with glycophytic (i.e., non-halophytic) dune species such as *Panicum turgidum*, *Crotalaria persica*, *Lotus garcinii*, *Taverniera spartea* and *Indigofera intricata*. Similar also is the halophytic vegetation of Qatar, which along a transect from the mangrove zone to the sabkha plain, show a distinct floristic and edaphic gradient with the following zonation: 1) *Avicennia marina*, 2) *Arthrocnemum glaucum*, 3) *Halocnemum strobilaceum*, 4) *Juncus rigidus–Aeluropus lagopoides*. Associated species are *Zygophyllum qatarense*, *Halopeplis perfoliata* and *Anabasis setifera*.

In Kuwait *Salicornia europaea* grows on low, frequently inundated mud banks or along creeks, sometimes associated with *Aeluropus lagopoides* and *Bienertia cycloptera,* or with *Juncus rigidus* on the fringes of creeks. A *Halocnemum strobilaceum* community occupies the lower marshes along the shoreline with the seaward edge inundated very frequently by tides. A *Seidlitzia rosmarinus* community occurs further inland, followed by *Nitraria retusa* above the high tide mark dominating the middle marshes, and finally the *Zygophyllum qatarense* community on elevated, coarse sandy sites on the landward edge of the marsh. The salt marshes are fringed by non-halophytic communities such as the *Cyperus conglomeratus* community, the *Rhanterium epapposum–Convolvulus oxyphyllus–Stipagrostis plumosa* community and the *Haloxylon salicornicum* community, the latter covering most of the territory of Kuwait.

*Acknowledgements*

My greatest thanks to the organising committee of the Biosaline Agriculture and High Salinity Tolerance Conference, the Rector and Vice Rector of Mugla University, Turkey for their invitation, and for their wonderful hospitality. My attendance to this Conference and research facilities were supported by the Royal Botanic Gardens Kew, which is gratefully acknowledged.

## References

1 Bogdan AV (1958) Some edaphic vegetational types at Kiboto, Kenya. *J Ecology* 46: 115–116
2 Vesey-FitzGerald DF (1963) Central African grasslands. *J Ecology* 51: 243–273
3 Gillett JB, Polhill RM, Verdcourt B (1971) Leguminosae. In: RM Polhill (ed.) In: *Flora of Tropical East Africa* 253. Crown Agents, London
4 Knapp R (1973) *Die Vegetation von Afrika*. Fisher, Stuttgart
5 Clayton WD, Phillips SM, Renvoize SA (1974) Gramineae. In: E Milne-Redhead, RM Polhill (eds) In: *Flora of Tropical East Africa* 2: 281–282. Crown Agents, London

6 Lind EM, Morrison MES (1974) *East African Vegetation*. Longman, London
7 White F (1983) The vegetation of Africa. A descriptive memoir to accompany the UNESCO, AETFAT, UNSO vegetation map of Africa. UNESCO, Paris
8 Ghazanfar SA, Fisher M (1998) *Vegetation of the Arabian Peninsula*. Kluwer Academic, The Netherlands
9 Deil U (1998) Coastal and sabkha vegetation. In: SA Ghazanfar, M Fisher (eds) In: *Vegetation of the Arabian Peninsula* 209–227. Kluwer Academic, The Netherlands
10 Mandaville JP (1998) Vegetation of the sands. In: SA Ghazanfar, M Fisher (eds) In: *Vegetation of the Arabian Peninsula* 191–206. Kluwer Academic, The Netherlands
11 Ghazanfar SA (1998a) Vegetation of the plains. In: SA Ghazanfar, M Fisher (eds) In: *Vegetation of the Arabian Peninsula* 175–190. Kluwer Academic, The Netherlands
12 Ghazanfar SA (1998b) Water vegetation. In: SA Ghazanfar, M Fisher (eds) In: *Vegetation of the Arabian Peninsula* 175–190. Kluwer Academic, The Netherlands
13 Phillips S (1997) Gramineae. In: I Hedberg, S Edwards (eds) In: *Flora of Ethiopia* 108–110. The National Herbarium, Biology Department, Addis Ababa University, Ethiopia and The Department of Systematic Botany, Uppsala University, Sweden
14 Thulin M (1993) *Flora of Somali* Vol. 1. Royal Botanic Gardens Kew
15 Danin A (1983) *Desert Vegetation of Israel and Sinai*. Cana Publishing House, Jerusalem
16 Frey W, Kürschner H, Stichler W (1985) Photosynthetic pathways and ecological distribution of halophytes from four littoral salt marshes (Egypt/Sinai, Saudi Arabia, Oman and Iran). *Flora* 177: 127–134
17 Mahmoud A, El-Sheikh AM, Youssef MM, Al-Tom M (1985) Ecology of the littoral salt marsh vegetation at al-Magawah, on the Gulf of Aqaba, Saudi Arabia. *Arab Gulf J Sci Res* 3: 145–163
18 El-Shourbagy MN, Al-Eidaros OH, Al-Zahrani HS (1987) Distribution of *Halopeplis perfoliata* (Forssk.) Bge. ex Schweinf. in the Red Sea coastal salt marshes: phytosociological relations and responses to soil. *J Coastal Res* 3: 179–187
19 El-Demerdash MA, Zilay AM (1994) An introduction to the plant ecology of Tihama plains of Jizan region, Saudi Arabia. *Arab Gulf J Sci Res* 12: 285–299
20 Al-Gifri AN, Hussein MA (1993) Plant communities between Abyan and Modia (Abyan Governorate, Jemen). *Feddes Repertorium* 104: 267–270
21 Al-Gifri AN, Al-Subai MJ (1994) Vegetation between Abyan and Modia (Abyan Governorate, Jemen). *Feddes Repertorium* 105: 229–234
22 Kürschner H, Al-Gifri AN, Al-Subai MY, Rowaished AK (1998) Vegetation pattern within coastal saline of southern Yemen. *Feddes Repertorium* 109:147–159
23 Ghazanfar SA (1993) Vegetation of the Khawrs and Fresh Water Springs of Dhofar. Part E in: *Khawrs and Springs of the Dhofar Governorate*. Survey and Monitoring Studies. Unpublished report. Planning Committee for Development and Environment in the Governorate of Dhofar, Oman
24 Ghazanfar SA (1995) Coastal sabkhas: an analysis of the vegetation of Barr al Hikman. In: A Khan, IA Ungar (eds) In: *The Biology of Salt Tolerant Plants* 277–283. Department of Botany, University of Karachi, Pakistan
25 Ghazanfar SA (1996) Analysis of the vegetation of saline, brackish and freshwater pools in Dhofar, Sultanate of Oman. In M Öztürk, Ö Seçmen, G Gork (eds) In: *Plant Life in Southwest and Central Asia* 501–508. Ege University Press, Izmir, Turkey
26 Ghazanfar SA (1999) Coastal Vegetation of Oman. *Estuarine, Coastal and Shelf Science* 49: 21–27
27 Ghazanfar SA, Rappenhöner D (1994) Vegetation and flora of the Islands of Masirah and Shagaf, Sultanate of Oman. *Arab Gulf J Sci Res* 12: 509–524

28 Frey W, Kürschner H (1986) Masqat Area (Oman), remnants of vegetation in an urban habitat. In: H Kürschner (ed.) In: *Contributions to the Vegetation of Southwest Asia* 201–221. Beihefte Tübinger Atlas Vorderen Orients, Reihe A (Naturwissenschaften) Nr. 24. Dr Ludwig Reichert Verlag, Wiesbaden

29 Kürschner H (1986a) A study of the vegetation of the Qurm Nature Reserve, Muscat area, Oman. *Arab Gulf J Sci Res* 4: 23–52

30 Deil U, Müller-Hohenstein K (1996) An outline of the vegetation of Dubai (UAE). *Verhandlungen der Gesellschaft für Ökologie* 25: 77–95

31 Batanouny KH (1981) *Ecology and Flora of Qatar*. Qatar University Press, Qatar

32 Babikir AA (1984) Vegetation and environment on the coastal sand, dunes and playas of Khor El-Odaid Area, Qatar. *Geo Journal* 9: 377–385

33 Babikir AA (1986) The vegetation of natural depressions in Qatar in relation to climate and soil. *J Arid Environments* 10: 165–173

34 Batanouny KH, Turki AA (1983) Vegetation of south-western Qatar. *Arab Gulf J Sci Res* 1: 5–19

35 Abdel-Razik MS, Ismail AM (1990) Vegetation composition of a maritime salt marsh in Qatar in relation to edaphic factors. *J Vegetation Sci* 1: 85–88

36 Abdel-Razik MS (1991) Population structure and ecological performance of the mangrove *Avicennia marina* (Forssk.) Vierh. on the Arabian Gulf coast of Qatar. *J Arid Environments* 20: 331–338

37 Babikir AA, Kürschner H (1992) Vegetational patterns within a coastal saline of NE-Qatar. *Arab Gulf J Sci Res* 10: 61–75

38 Abbas JA, Saleh MA, Mohammed SA (1991a) Edaphic factors and plant species distribution in a protected area in the desert of Bahrain Island. *Vegetatio* 95: 87–93

39 Abbas JA, Saleh MA, Mohammed SA (1991b) Plant communities of a protected area in the desert of Bahrain island. *J Arid Environments* 20: 21–42

40 Abbas JA, El-Oqlah AA (1992) Distribution and communities of halophytic plants in Bahrain. *J Arid Environments* 22: 205–218

41 Halwagy R, Halwagy M (1977) Ecological studies on the desert of Kuwait. III. The vegetation of the coastal salt marshes. *J Uni Kuwait (Science)* 4: 33–73

42 Halwagy R, Moustafa AF, Kamal S (1982) On the ecology of the desert vegetation in Kuwait. *J Arid Environments* 5: 95–107

43 Halwagy R (1986) On the ecology and vegetation of Kuwait. In: H Kürschner (ed.) In: *Contributions to the Vegetation of Southwest Asia* 81–109. Beihefte zum Tübinger Atlas des Vorderen Orients, Reihe A (Naturwissenschaften) Nr. 24. Dr Ludwig Reichert Verlag, Wiesbaden

44 Kürschner H (1986b) Omanisch-makranische Disjunktionen Ein Beitrag zur pflanzengeographischen Stellung und zu den florengenetischen Beziehungen Omans. *Botanische Jahrbücher für Systematik* 106: 541–562

45 Knapp R (1968) Höhere Vegetationseinheiten von Äthiopien, Somalia, Natal, Transvaal, Kapland und einigen Nachbargebieten. *Geobotanische Mitteilungen* 56: 1–36

46 Lieth H, Berlekamp J, Fuest S, Riediger S (1999) *Climate Diagram World Atlas: CD-ROM 1, Climate and Biosphere*. Backhuys Publishers, Leiden

Biosaline Agriculture and Salinity Tolerance in Plants
Edited by M. Öztürk, Y. Waisel, M.A. Khan and G. Görk


# Vegetation zones in the salty marshes of Central Anatolia and natural borders of agricultural usage (Turkey)

Ahmet Aksoy[1] and Ergin Hamzaoğlu[2]

[1] *University of Erciyes, Science & Art Faculty, Dept. of Biology, 38039 Kayseri, Turkey*
[2] *University of Erciyes, Science & Art Faculty of Yozgat, Dept. of Biology, Yozgat, Turkey*

## Introduction

The region of Central Anatolia is rich in terms of lakes. These, Tuz Lake (Konya–Ankara–Aksaray), Seyfe Lake (Kırşehir) and Sultansazlığı (Kayseri) are surrounded by salty marshes. In the arid areas around these lakes, there are halophytic vegetation zones with changing physiognomy and floristic composition. In this study, edafic factors that cause the formation of vegetation zones, floristic composition of the zones and their situation from the point of view of agricultural usage are evaluated.

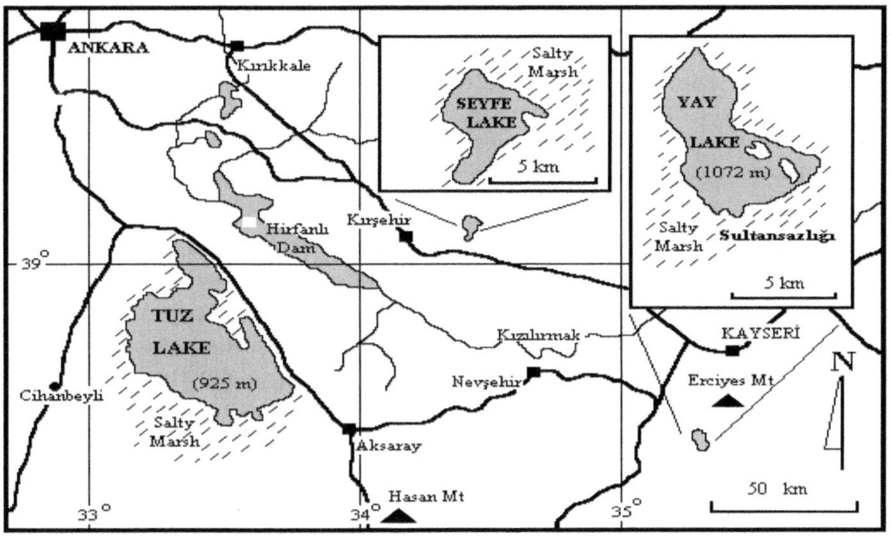


**Figure 1.** The most important salty marshes of Central Anatolia; Tuz Lake, Seyfe Lake and Sultansazlığı

Tuz Lake (925 m) is the biggest lake of Central Anatolia and is situated within the city borders of Ankara, Aksaray and Konya. Seyfe Lake (1,110 m) is in Kırşehir and Sultansazlığı is in the city borders of Kayseri (Fig. 1).

These lakes, which occupy the plains where drainage is spoilt in the region are under the influence of semi-arid or arid, very cold Mediterranean climate (Tab. 1). Annual average temperature changes range between 11.1–11.9° C. The highest temperatures are seen in July and August when the rain is minimum. The arid season, which begins at the end of June, continues for 4–5 months [1–3].

**Table 1.** The climate data of the closest meteorological stations to Tuz Lake, Sultansazlığı and Seyfe Lake

| Areas and Stations | Altitude (m) | P (mm) | M (° C) | m (° C) | Q | PE (mm) | S | Preciptitat regime | Bio-climate |
|---|---|---|---|---|---|---|---|---|---|
| Tuz Lake (Aksaray) | 980 | 347.4 | 299 | 13.4 | 36.4 | 36.5 | 1.2 | Sp-W-A-S | Semi-arid, lower-very cold Mediterranean climate |
| Tuz Lake (Cihanbeyli) | 969 | 308.0 | 296 | −3.8 | 32.7 | 33.1 | 1.1 | Sp-W-A-S | Arid, upper-very cold Mediterranean climate |
| Sultansazlığı (Develi) | 1180 | 366.6 | 299 | −4.4 | 37.4 | 31.2 | 1.0 | Sp-W-A-S | Semi-arid, lower-very cold Mediterranean climate |
| Sultansazlığı (Yeşilhisar) | 1150 | 274.8 | 309 | −4.3 | 27.2 | 40.0 | 1.2 | Sp-W-A-S | Arid, upper-very cold Mediterranean climate |
| Seyfe Lake (Kırşehir) | 1100 | 351.5 | 352 | −3.6 | 31.3 | 36.2 | 1 | Sp-W-A-S | Arid, upper-very cold Mediterranean climate |

P (mm): Mean annual precipitation, M (° C): Mean maximum for the hottest month, m (° C): Mean minimum for the coldest month, Q: Emberger's pluviometric quotient (2000. P/$M^2 - m^2$), PE: Summer rainfall, S: Emberger's index of xericity (S=PE/M), W: Winter, Sp: Spring, S: Summer, A: Autumn

The region of Central Anatolia forms the shape of a bowl, surrounded by mountains. Owing to its isolated nature, the region has a characteristic plant cover. In terms of phytogeography, it is included in the Irano-Turanian region and it forms a province called 'Central Anatolian' [4]. The salty marshes in the region constitute the richest areas of this province in terms of endemism (especially Tuz Lake). The first studies of the terrestrial salty marshes of Central Anatolia are related to halophytic communities of Tuz Lake and Konya plain [5, 6]. Thanks to the detailed phytosociological studies that have been realized in the last years, syntaxonomy of halophytic vegetation of Central Anatolia has been determined [7–10].

The region of Central Anatolia has a favorable topography for agriculture. This situation causes the region to be under intensive anthropogenic influence. Because of the fast increase in population increase and technological developments during the last 60 years, many pasture regions have been converted into farmland. Salty marshes have suffered most from these activities. *Artemisia santonicum* L. communities that constitute the most outer zone of these areas are known to have spread in much larger areas before the agricultural activities became dense [11].

Recently, it has been seen that salty marshes are very rich in the diversity of species and it should be protected more effectively. However, it is not easy to draw borders between the areas that will be protected and that will be opened to agriculture. In the shaping of the borders, economical and political desires of human bodies are more diagnostic. Consequently, the borders of protection areas are kept as limited as possible. As the floristic and faunistic richness decreases over time, the faults that were made have been noticed and broadening the borders in many areas that have been protected today is an obligation.

## Materials and methods

The results of the analyses of the soil samples that belong to the zones identified in the base of Yay Lake (Sultansazlığı, Kayseri), phytosociological studies made on salty marshes of Central Anatolia and observation of the authors constitute the material of this study. The soil samples were collected on 29 August 2004 and dried in shadow and sieved by a 2 mm sieve. Solving and reading processes are made according to [12].

## Results

In the salty marshes, vegetation is formed by mixed zones. The most important factor shaping these zones is soil salinity and the salt tolerance limits of species. It means that the existence of the species in any zone is not a coincidence. Therefore, floristic composition of any zone gives us very important information about formations of the edaphic properties of that zone.

The zone formation of halophytic species, depending on their tolerance limits, is a basic property that can be observed from their first development. Although this is already known, it will be useful to show it again here. Yay Lake is the biggest lake in Sultansazlığı (Kayseri). The water of Yahyalı and Kovalı streams which feed this lake is accumulated in dams and used for agriculture. The amount of water in the lake has decreased over the last 5 years as the result of intensive agricultural activities. In the drying part of the lake, primary development of the vegetation (= halosere) has started. These areas are in the aggregation phase and have three zones. Each zone consists of single and dominant species. Although sometimes different species enter the zones, this situation doesn't affect physiognomy. The dominant species that form the zones in halosere, and analysis results of soil samples taken from these zones, are given in Table 2. According to results, the main cations and anions that cause the salinity of the soil are $Mg^{2+}$, $Na^{+}$, $Ca^{2+}$, $K^{+}$, $Cl^{-}$, $SO_4{}^{2-}$ and total $CO_3{}^{2-}$. As the amount of these increase, salinity increases. In Yay Lake, *Salicornia europaea* L. takes place in the first zone where these factors are in the highest density. These values decrease gradually in the second and third zones. Depending on this, there is

*Salsola macera* Litw. in the second zone and *Petrosimonia nigdeensis* Aellen in the third zone. These findings reveal the basic principals of simple but necessary zone formation.

**Table 2.** Soil analyzes results belonging to primary vegetation zones developed in the base of Yay Lake

| Zones | Dominant species | pH ▼ | Cations ($\mu gg^{-1}$) Mg$^{2+}$ ▼ | Na$^{+}$ ▼ | Ca$^{2+}$ ▼ | K$^{+}$ ▼ | Anions ($mgg^{-1}$) Total $CO_3^{2-}$ ▼ | $Cl^{-}$ ▼ | $SO_4^{2-}$ ▼ |
|---|---|---|---|---|---|---|---|---|---|
| 1 | *Salicornia europaea* | 8.43 | 4,415 | 4,349 | 124,800 | 4,265 | 453 | 8.40 | 2.85 |
| 2 | *Salsola macera* | 8.35 | 4,304 | 3,343 | 64,438 | 3,999 | 428 | 5.29 | 1.89 |
| 3 | *Petrosimonia nigdeensis* | 8.20 | 4,282 | 2,406 | 55,384 | 3,359 | 414 | 3.34 | 1.42 |

The beginning of the climax vegetation seen in Central Anatolia is no doubt similar to the example of Yay Lake. These zones of vegetation have different characteristics not only in terms of physiognomy but also floristic composition and endemism.

*1. Zone*

In the inner zone where the salt density is maximum in Tuz Lake, Seyfe Lake and Sultansazlığı, in the first zone, *Salicornia europaea* L. and *Halocnemum strobilaceum* (Pall.) M.Bieb. are dominant (Fig. 2). The general cover in the zone changes between 10–80 % depending on salt density. The width differs according to the slope and the salt concentration. The zone is rather narrow in the regions with high slopes, but it can continue kilometers in the regions with very small slopes. Besides, this zone is observed in more hollow and, therefore, saltier areas in the other zones. Because of its high salt concentration, it is quite poor in terms of floristic composition. The number of species usually changes between 1 and 10 and it doesn't have any endemic

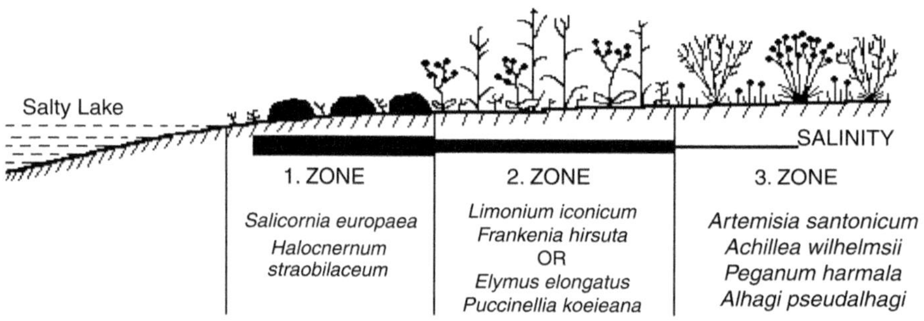


**Figure 2.** Sequence of vegetation zones depending on salinity in Tuz Lake, Seyfe Lake and Sultansazlığı

species. *Halocnemum strobilaceum* can also be seen in the second zone where the salt concentration decreased a little. In this situation floristic composition becomes a little richer and it can contain a few endemic species [5, 7–9, 13, 14].

*2. Zone*

Depending on the gradually decreasing salinity in the second zone, an instantaneous increase is observed in the number of species. In Central Anatolia this zone is seen with two different mixed physiognomies. The first physiognomy is dominant in the areas where water sources are not available. In this physiognomy where short succulent chaemophytes and hemicryptophytes are dominant, the number of species changes between 65–80 and endemism between 16–21 %. The most common species are *Limonium iconicum* (Boiss. and Heldr.) O. Kuntze, *Frankenia hirsuta* L., *Lepidium caespitosum* Desv., *Halimione verrucifera* (M.Bieb.) Aellen, *Salsola stenoptera* Wagenitz and *Puccinellia convoluta* (Hornem) P.Four, and *Halocnemum strobilaceum* which is also available in the first zone (Fig. 2). From these species, *Limonium iconicum*, *Lepidium caespitosum* and *Salsola stenoptera* are endemic for the salty marshes of Central Anatolia. In the areas where the water sources that feed the marsh are available, another physiognomy is dominant which is on the shape of tall herbaceous hemicryptophytes. Dominant species are *Elymus elongatus* (Host) Runemark subsp. *salsus* Melderis, *Puccinellia koeieana* Melderis subsp. *anatolica* Kit Tan, *Inula aucherana* DC., *Juncus maritimus* Lam., *Juncus heldreichianus* Marsson ex. Parl. subsp. *orientalis* Snog. and *Puccinellia convoluta* which is also available in the first zone. *Elymus elongatus* subsp. *salsus* is endemic for the salty marshes of Central Anatolia. In these areas the number of species changes between 45–60 and endemism between 26–32%. *Chenopodiaceae* and *Plumbaginaceae* are dominant in the first physiognomy and *Poaceae* and *Juncaceae* are in the other [5, 7–9, 14].

*3. Zone (salty steppe)*

In the salty marshes of Central Anatolia, the most outer zone is characterized by *Artemisia santonicum* L. (Fig. 2). This zone, called 'salty steppe', forms a border with the non-halophytic communities around. Succulent species are not seen very often in the zone. The individual cover of *Artemisia santonicum* is sometimes 70 %. The other common species of the zone are *Peganum harmala* L., *Alhagi pseudalhagi* (M.Bieb.) Desv., *Achillea wilhelmsii* K.Koch, *Noaea mucronata* (Forssk.) Aschers. and Schweinf. subsp. *mucronata* and *Apera intermedia* Hackel. In this zone the number of species changes between 70–95 and endemism rate between 17–23 % [5, 7, 10, 13].

In the first and second vegetation zones, usually true halophytes that have adapted to high salt concentrations grow. The vegetation period begins fairly late because these parts of the marsh are under water for a long part of the year. The upper surface starts to dry towards the end of July and it gets hard enough to be stepped on. In the salty steppes, vegetation period begins almost at the same time as the non-halophytic communities around. In this community, which is never under water in any period of the year, there are usually no true halophytes. Non-halophytic species with a wide tolerance to salt and miohalophytes that can survive in a little salty environment are common in the zone.

The vegetation zones were compared in terms of floristic similarity by utilizing the phytosociological studies in the steppe areas in Tuz Lake and in the vicinity of Cihanbeyli which is a border to this area [9, 10, 15, 16]. According to this comparison, there is no floristic similarity between the first vegetation zone of Tuz Lake and Cihanbeyli steppe vegetation. In the second zone there is about 5 % floristic similarity in the areas where water sources are available and about 14 % in the areas where water sources are not available. Floristic similarity between salty steppes and Cihanbeyli steppe vegetation is about 38 %. These ratios show that the similarity to non-halophytic vegetation in the vicinity increases as the salinity decreases.

*Artemisia santonicum* communities shape the last zone in almost all salty marshes of Central Anatolia. They form a 'natural border' between true halophytic and non-halophytic communities. A big part of salty steppes have been converted into agricultural areas. In addition, the second and first zones have tried to be converted to agricultural areas by constructing drainage canals in some areas.

## Discussion

Central Anatolia Region is the richest region of Turkey in terms of terrestrial salty marshes owing to its topographic structure. Out of these marshes, Tuz Lake is declared as a 'Special Environment Protection Area' and Sultansazlığı as a 'Nature Protection Area' and they are under protection. At the same time, these three areas are among the important 'watery areas' because they have a rich flora and they are the home of many bird species. Turkish Republic undertook to protect these areas by signing international treaties like RAMSAR (1971), BERN (1979) and RIO (1992).

However, the situation is not so positive today. In these areas, almost the entire salty steppe zone has been converted into agricultural areas. The other vegetation zones are under heavy threat. Recently, some salty marshes were almost killed by making new drainage canals or using water sources. When overgrazing, illegal hunting and collecting some plant species systematically for commercial purposes are added to the others activities. Hence, the end seems to be very close. One of the important dangers for the future of these areas is that people are not aware of the richness they have. Mentioning the salty marshes as 'deserts' in many regions shows the point of view about these areas.

Salinity is a serious threat in terms of productivity of the soil and it must be struggled with necessarily [17]. But when struggling with salinity, natural richness must be protected at the same time. Struggling with salinity in the present agricultural areas and trying to dry salty marshes which are quite rich in terms of biodiversity is difficult.

Today agricultural technology is quite developed and more products can be taken from a unit of area. These advances show that it is enough to make the present agricultural areas more productive instead of opening new agriculture areas. Struggling with salinity must be intensified in the regions like Çukurova, Harran, Muş and Konya plains that have large agricultural areas.

Salty marshes should not be evaluated as the areas to struggle with salinity, but as the areas to be protected. When we look at the previous applications, the protection borders of salty marshes are usually determined as the second zones and even in some places, the first zones. Yet, first and second zones are the areas that must definitely be protected. Therefore, the protection borders in the salty marshes that are under protection in Central Anatolia should be reviewed again. Borders must begin from the salty marshes and if possible, the salty steppe zone with an apparent width must be also included in this border.

Today we should think and act more differently than we have in the past. We should not wait to lose the richness that we have in order to understand our faults. We must see this richness as a deposit and we should protect them with attention and we should try to transfer them to the future.

*Acknowledgements*
We are grateful to Asst. Prof. Dr. Uğur Şahin, one of the teaching staff of E.U. Chemistry Department for his help in analyzing the soil samples.

## References

1 DMI (1984) *Ortalama, Extrem Sıcaklık ve Yağış Değerleri Bülteni* (Günlük-Aylık). Devlet Meteoroloji İşleri, Ankara
2 Akman Y, Daget PH (1971) Quelques aspects synoptigues des climats de la Turquie. *Bulletin Societe Languedocienne Geographie* 5: 270–300
3 Akman Y (1982) Climats et bioclimats mediterranéens en Turquie. *Ecologia Mediterranea* VIII (1, 2): 73–87
4 Zohary M (1973) *Geobotanical Foundations in the Middle East*. Vols 1–2, Gustav Fischer Verlag, Stuttgart
5 Birand H (1961) Orta Anadolu bozkırında vejetasyon incelemelerinin ilk sonuçları. I. Tuz Gölü çorakçıl bitki birlikleri. *Topraksu Umum Müdürlüğü Neşriyatı* 103: 1–56
6 Birand H (1970) Die Verwüstung der Artemisia-Steppe Bei Karapınar in Zentralanatolien. *Vegetation* 20: 21–47
7 Yurdakulol E (1974) Konya ovasındaki çorak bataklıkların vejetasyonunun bitki sosyolojisi yönünden araştırılması. *Bitki* 1: 257–277
8 Yurdakulol E, Ercoşkun T (1990) Orta Anadolu'da çorak alanların vejetasyonu üzerinde ekolojik ve sintaksonomik araştırma. *Turkısh J Botany* 14: 109–123
9 Aydoğdu M, Hamzaoğlu E, Kurt L (2002) New halophytic syntaxa from Central Anatolia (Turkey). *Israel J Plant Sci* 50: 313–323
10 Aydoğdu M, Kurt L, Hamzaoğlu E, Ketenoğlu O, Cansaran A (2004) Phytosociological studies on salty steppe communities of Central Anatolia, Turkey. *Israel J Plant Sci* 52: 71–79
11 Birand H (1947) *Über die Vegetationsverhältnisse der Artemisia-Steppe als Weide*. Ann. Université d' Ankara I
12 Allen ES (1989) *Chemical analysis of ecological materials*. Blackwell Scientific Publications, Oxford

13 Yurdakulol E, Öncel I, Demirörs M, Yıldız A, Keleş Y (1996) Ecological and syntaxonomic investigation of salt marshes vegetation in of Burdur and Acıgöl (Denizli-Turkey). *Ecologia Mediterranea* 22: 51–61
14 Davis PH (1965–1985). *Flora of Turkey and the East Aegean Islands*. Vol. 1–9: 1965 (Vol. 1), 1967 (Vol. 2), 1970 (Vol. 3), 1972 (Vol. 4), 1975 (Vol. 5), 1978 (Vol. 6), 1982 (Vol. 7), 1984 (Vol. 8), 1985 (Vol. 9), Edinburgh
15 Sorensen TA (1948) A method of establishing groups of equal amplitude in plant sociology based on similarity of species content. *Biol Skr K d Vidensk Selsk* 5: 1–34
16 Hamzaoğlu E, Aydoğdu M, Kurt L, Cansaran A (2004) New syntaxa from the West Part of Central Anatolia. *Pakistan J Botany* 36: 235–246
17 Oakes H (1958) Türkiye Toprakları. Türk Ziraat Mühendisleri Birliği Neşriyatı, Ankara

Biosaline Agriculture and Salinity Tolerance in Plants
Edited by M. Öztürk, Y. Waisel, M.A. Khan and G. Görk


# Halophytes as cash crops for animal feeds in arid and semi-arid regions

Hassan M. El Shaer

*Animal and Poultry Production Division, Desert Research Center, P.O. Box 11753, Mataria, Cairo, Egypt*

## Introduction

Dry regions occupy more than one-third of the Earth's land area and are inhabited by approximately 16% of the World's population. In the arid countries, population increase, global climate change, a large number of the nomadic populations as well as over-use of renewable resources such as water has disturbed the sensitive ecological balance. Water resources in many of these countries are dwindling both in quantity and in quality. Such countries suffer from severe shortage of freshwater because of the dry climate and the irregularity of rainfall. Moreover, water resources of large parts of several countries are saline. The soils there are either naturally saline or have been salinized, mainly because of ill-conducted irrigation [1]. The potential utilization of halophytes grown under saline water irrigation would conserve freshwater and will enable crop production in marginal areas [2]. This article aims to draw attention to possible utilization of halophytes as cash crops, particularly in dry and saline ecosystems, and to focus primarily on the potential use of halophytes as animal fodder in arid and semi-arid regions.

## Possibilities for halophytes utilization

There are more than 2,000 plant species registered as halophytes [3]. Several halophytes are presently used along seashores and inland salt marshes extensively. In salt affected areas, *Puccinellia* (*Puccinellia ciliata*), tall wheat grass (*Thinopyrum ponticum*) and Chenopods (e.g., various species of *Atriplex*, *Kochia* and *Bassia*) have been planted for the improvement of soil conditions due to their low transpiration rate, high efficiency of water utilization, drought and salinity resistance [4]. They get increasing attention today because of the steady increase of the salinity in irrigation systems in Mediterranean and subtropical desert countries where the increasing population has reached the limits of freshwater availability. Many halophytes have the potential to become cash crops for several reasons [2]:

- Halophytes for food such as *Aster tripolium* (salt aster), *Salicornia* sp., *Avicennia marina* and *A. germinant.*
- Halophytes for fodder such as many species of *Atriplex*, *Tamarix*, *Nitraria retusa* and grasses. All can provide good fodder for livestock and wildlife.
- Halophytes for wood such as *Tamarix* spp, and mangroves.
- Halophytes grown for chemicals: A variety of halophytes are collected for health and beauty products purposes or for tanning [3].
- Halophytes for landscaping: Fast growing plants can cover barren soils in a short time, e.g., *Batis maritima*, *Sesuvium portulacastrum* and *Atriplex* spp.
- Ornamental halophytes such as *Limoniastrum monopetalum* and *Aster tripolium.*
- Industrial raw materials: Growing halophytes for biomass is economically feasible only if additional elements, for compounds with especial value, are involved.
- Environmental protection: Many halophytic species are used for coastline protection such as *Spartina alterniflora*, *Spartina maritima* and *Avicennia marina.*

A number of considerations should be taken into account to assess the sustainability of a halophyte utilization system. The choice of the suitable crop depends on its local usefulness, the climatic zone, the adapted techniques for agricultural production and the ultimate objectives. The relevant aspects should be considered with the most important factor being water availability. However, economical prospects for halophytes can be evaluated when they are looked upon beyond their salt tolerance values. There are a number of other factors related to the plants establishment, e.g., productivity, water requirements, forage quality, etc., which could qualify the species for agricultural purposes. The following section briefly discusses the potential role of halophytes as feed materials.

## Halophytes as feed resources

Shortage of feed resources is a common characteristic in arid and semi-arid regions and is considered the main constraint to improving livestock productivity. Therefore, intensive efforts have been directed to find alternative feed resources from saline habitats. Halophytes that are used as forage species will have better cash values if they have better forage quality. High palatability, digestibility and good nutritional value (high protein and lesser fiber, ash and oxalate contents) would significantly improve the forage quality.

## Nutritive value of common halophytes

The biomass production, palatability and nutritive value of halophytes vary from area to area within the same region and from season to season depending upon several environmental factors [5]. Moreover, palatability and nutritive values vary significantly among halophyte species (Tab. 1). In general, most halophytes contain sufficient levels of crude protein, other essential nutrients that seems to cover the nutritional requirements of animals, particularly during the wet season [6, 7]. Fibrous

**Table 1.** Average value of chemical composition* and palatability index (PI)** of common halophytes (% on DM basis). [Sources: 6, 14]

| Plant species | DM | CP | CF | EE | Ash | NEF | PI | | |
|---|---|---|---|---|---|---|---|---|---|
| | | | | | | | Goats | Sheep | Camels |
| ***Halocn.strobil.ceum*** | | | | | | | | | |
| wet season | 28.6 | 6.78 | 14.6 | 2.46 | 35.7 | 40.46 | NP | NP | PP |
| dry season | 37.8 | 4.22 | 19.6 | 2.16 | 42.5 | 31.52 | NP | PP | PP |
| ***Zygophyllum album*** | | | | | | | | | |
| wet season | 30.4 | 7.12 | 14.6 | 2.25 | 26.5 | 49.42 | NP | NP | PP |
| dry season | 39.3 | 6.30 | 16.2 | 1.63 | 28.5 | 47.31 | NP | NP | PP |
| ***Tamarix mannifera*** | | | | | | | | | |
| wet season | 57.4 | 7.64 | 16.1 | 2.21 | 26.0 | 47.97 | PP | PP | FP |
| dry season | 63.1 | 6.26 | 17.5 | 1.78 | 30.9 | 43.47 | PP | NP | FP |
| ***Juncus acutus*** | | | | | | | | | |
| wet season | 35.0 | 7.11 | 28.5 | 2.35 | 12.3 | 49.94 | FP | FP | FP |
| dry season | 42.3 | 6.00 | 33.5 | 2.05 | 14.0 | | PP | PP | FP |
| ***Salsola tetrandra*** | | | | | | | | | |
| wet season | 45.0 | 9.73 | 12.4 | 2.61 | 30.1 | 45.10 | FP | FP | HP |
| dry scason | 49.9 | 8.38 | 14.4 | 1.67 | 35.8 | 39.68 | PP | PP | HP |
| ***Nitraria retusa*** | | | | | | | | | |
| wet season | 14.6 | 9.10 | 12.8 | 3.01 | 16.2 | 58.87 | HP | HP | HP |
| dry season | 20.1 | 7.20 | 18.2 | 2.28 | 19.7 | 52.62 | PP | FP | HP |
| ***Atriplex halimus*** | | | | | | | | | |
| wet season | 42.1 | 9.21 | 18.7 | 3.20 | 31.4 | 37.49 | FP | FP | HP |
| dry season | 58.4 | 6.32 | 22.6 | 3.10 | 36.7 | 31.28 | PP | PP | FP |
| ***Atriplex nummularia*** | | | | | | | | | |
| wet season | 21.7 | 13.3 | 24.2 | 5.09 | 21.8 | 34.98 | HP | HP | HP |
| dry season | 39.8 | 10.2 | 28.0 | 1.90 | 26.7 | 33.81 | HP | HP | HP |
| ***Suaeda fruticosa*** | | | | | | | | | |
| wet season | 30.9 | 11.10 | 10.9 | 3.90 | 25.4 | 48.66 | HP | HP | HP |
| dry season | 48.3 | 8.40 | 13.7 | 3.00 | 30.3 | 44.60 | FP | HP | HP |

*DM: dry matter; CP: crude protein; CF: crude fiber; EE: ether extract, NFE: nitrogen free extract, **Palatability index : HP = Highly palatable, FP = Fairly palatable, PP = Poorly palatable

materials and ash contents are higher and increase while gross energy and protein contents are low and decrease with advancing maturity of the plants [8]. Therefore, most halophytes are nutritious during wet seasons and can sustain the maintenance requirements of animals. However, they are poor in summer and autumn (dry season) and need to be supplemented with other feed ingredients, particularly with energy feed resources [9, 10]. The chemical composition of *Atriplex* species, their nutritive values (digestible crude protein, DCP) and the total digestible nutrients (TDN) varied greatly [11, 12]. Salt marsh plants appear to contain sufficient concentrations of most major minerals (Tab. 2). They have no harmful effects even when high concentrations of such minerals are found in some plants [13]. However, some halophytes are deficient in sulfur and phosphorus [6]. No deficiency or toxicity of trace elements was observed for most of the halophytic species [14, 13]. From data in Tables 1 and 2, it appears that the high ash content of most halophytes, in particular *Atriplex* spp, leads to high salt intake and consequently to a high water intake (average of 7 liters/day/head of sheep) and high urine output [7]. The high salt content of many halophytes has been shown to reduce the apparent energetic value and increased the flow of shrub materials through the digestive system [15]. The ash content of *Atriplex* spp is, therefore, an additional reason for mixing the shrubby materials with a low ash feed. The inclusion of a supplement, particularly of a soluble carbohydrate source, to the diet of sheep or goats eating halophytes showed an increase in the utilization of such forages. This is because it increases intake, digestibilities, growth rate and other animal performances [16]. Therefore, available sources of energy, phosphorus and sulfur should be supplemented to animals fed halophytes, particularly during the critical physiological status of animals [6, 9]. Performance of sheep and goats fed with *Atriplex nummalaria* and supplemented with crashed barley grains during the four seasons of the year was determined [8] and data presented in Table 3. It seems that DM, water, TDN, and DCP intakes, digestibility, nitrogen retention and weight changes were significantly affected by the season. Both sheep and goats utilized efficiently the supplemented *Atriplex* during the winter and spring seasons which was positively reflected by their higher body weight changes and all other criteria of feed

**Table 2.** Overall averages of major elements in *Atriplex* spp (% of DW) [Source: 13]

| ***Atriplex* spp** | **Na** | **K** | **Ca** | **P** | **Cl** | **Mg** | **S** |
|---|---|---|---|---|---|---|---|
| *A. halimus* | 4.41 | 1.33 | 1.51 | 0.29 | – | 0.32 | 0.17 |
| *A. leucoclada* | 4.10 | 3.00 | 1.36 | 0.19 | 8.40 | 0.35 | 0.13 |
| *A. vesicaria* | 5.87 | 4.05 | 1.56 | 0.15 | 9.10 | 0.36 | – |
| *A. farinose* | 1.69 | 1.11 | 2.32 | – | 3.19 | 0.23 | 0.10 |
| *A. nummularia* | 4.15 | 3.19 | 1.26 | 0.11 | 7.80 | 0.30 | 0.20 |
| *A. canescens* | 3.90 | 2.08 | 1.43 | 0.16 | – | – | 0.18 |
| *A. semibaccata* | 3.32 | 1.35 | 1.29 | 0.11 | 0.10 | 0.33 | 0.25 |
| *A. glauca* | 6.30 | 1.60 | 2.14 | 0.13 | – | 0.56 | 0.11 |

**Table 3.** Utilization of supplemented *Atriplex nummularia* by sheep and goats [Source: 8]

| **Item** | Sheep | | | | Goats | | | | F test |
|---|---|---|---|---|---|---|---|---|---|
| | W1 | SP | SU | AU | W1 | SP | SU | AU | |
| BW changes, g/d | $55.6^b$ | $77.8^a$ | $23.2^c$ | $11.1^d$ | $49.4^b$ | $66.6^a$ | $17.1^c$ | $-9.1^d$ | ** |
| DM intake, g/d/Kg$^{0.75}$ | $52.0^b$ | $59.8^a$ | $47.7^{cd}$ | $45.8^{cd}$ | $53.7b^b$ | $58.0^a$ | $49.8^c$ | $44.6^c$ | ** |
| Water intake, ml/d/Kg$^{0.82}$ | $262^c$ | $311^b$ | $381^a$ | $356^a$ | $219^d$ | $257^c$ | $288^{bc}$ | $269^c$ | ** |
| N balance, mg/d/Kg$^{0.75}$ | $99^b$ | $143^a$ | $49^c$ | $-25^b$ | $86^b$ | $126^a$ | $39^c$ | $-24^b$ | ** |
| Digestibility, %DM | $60.1^a$ | $61.4^a$ | $58.1^b$ | $58.4^b$ | $58.4^b$ | $61^a$ | $59.2^b$ | $58.0^b$ | ** |
| TDN, g/d/Kg $^{0.75}$ | $24.0^b$ | $28.4^a$ | $21.2^{bc}$ | $19.0^c$ | $25^{ab}$ | $28.7^a$ | $22.0^{bc}$ | $18.7^c$ | ** |
| DCP , g/d/Kg $^{0.75}$ | $3.4^b$ | $5.0^a$ | $2.8^{bc}$ | $1.9^c$ | $3.5^b$ | $4.8^a$ | $3.0^{bc}$ | $1.9^c$ | ** |

W1, SP, SU, AU: winter, spring, summer and autumn, respectively, $^{**}P < 0.01$; Values on the same line with different superscript are significant utilization. The opposite was found for the dry season (summer and autumn). These results reconfirmed earlier findings that additional feed supplements are required particularly during the dry seasons [6].

## Enhancement of halophytes feed materials

Halophytes as individual forage materials have little prospect because long feeding periods are known to have adverse affects on browsing animals. Halophytic species vary in their palatability and acceptability to various animal species due to different factors [13]. Some of these plants can be fed to animals or grazed directly as fresh feed materials, particularly those that have high or moderate palatability and nutritious. Such plant species, i.e., *Atriplex* spp, *Nitraria retusa* and *Saudea fraticosa*, are always over grazed and disappear rapidly due to high grazing pressure of most animal species [6]. However, mixing halophyte forage with feed materials that are rich in protein or energy can significantly improve the feed nutritional value. A mix feed of dry grass and green *Atriplex* materials have been reported to increase the weight of goats in Pakistan [17]. Conversely, several halophytes are less palatable or unpalatable, producing a large biomass throughout the year. Utilizing such halophytes in arid countries is becoming necessary, particularly during the dry seasons or during prolonged periods of drought when other feed resources are scarce. These plants have some secondary metabolites, or so-called anti-nutritional factors, such as tannins, alkaloids, saponines, nitrites, etc., which negatively affect their use. Several approaches were undertaken to improve the utilization of such halophytic species through different processing treatments in order to improve their palatability and nutritive values [16, 18]. For instance, chopping can dramatically improve palatability of succulent species and enable a more efficient utilization of whole shrubs. Proper conservation such as hay making, haylage or ensiling processes of some halophytes could improve their utilization as good quality fodder. Results obtained by [16, 18, 19] for sheep and goats that were fed on various halophytic shrubs are summarized in Table 4. Different forms and combinations with other feed ingredients such as broiler litter, crushed date seeds, fodder beet and other forages were investigated. Ensiling process seems to improve the palatability of halophytes and their acceptability by sheep and goats. This might be due to the effect of anaerobic fermentation during the ensiling process on some anti- nutritional factors such as tannins and other phenolic compounds [20]. Additionally, the ash content and fiber materials are lowered [18]. Maximum intakes were improved greatly particularly for animals fed *Halocnemum strobilaceum* and *Atriplex halimus* silages (approximately 59 g DM/Kg$W^{0.75}$) followed closely by *Tamarix aphylla* (averaged 43 g DM/Kg$W^{0.75}$). Conversely, voluntary feed intake was increased by ensiling a mixture of some halophytic species, *A. halimus*, *H. strobilaceum*, *Tamarix mannifera* and *Zygophyllum album* with some agro-industrial byproducts such as ground date seeds, olive pulp, etc. [9]. Ensiling such halophytes with air-dried broiler litter improved their palatability and intake as compared to fresh or air-dried shrubs (Tab. 4). Animals fed these ensiled materials were in a positive nitrogen balance and tended to gain appreciable weight [18]. Similarly, [21] reported that sheep and goats consumed higher amounts of silage comprised of a mixture of *Acacia saligna* and *Atriplex nummularia*, as compared with fresh or air-dried materials. Such silage, as sole basal diets, provided sufficient digested nutrients (TDN and DCP) to meet the maintenance requirements for sheep and goats. Feeding these silages to animals seemed to be more economic since feed costs decreased (about 30–50%) in comparison with conventional diets such as berseem hay [16, 18]. Recently, sex silages were formulated from mixing different natural and cultivated halophytic shrubs with other feed ingredients [22]. It appeared that silage 5, which contained ensiled cultivated (*A. nummularia* and *Acacia saligna*) and natural shrubs (*Tamarix mannifera*, *Zygophyllum album* and *H. strobilaceum*) with broiler litter, showed the highest level of crude protein (13.9 %). The CP contents of the tested silages, in particular silage 5, seemed to be reasonable to maintain the protein requirements of ruminants. The same authors also indicated that all silages attained higher ash contents which could be attributed to the inclusion of the natural halophytic shrubs. These silages contained higher levels of ash, Na and K concentrations which appeared to be in normal ranges and enough to cover the requirements of ruminants without an adverse indication.

**Table 4.** Intake of some halophytes in fresh, air-dried and ensiled forms [Sources: 12, 13, 19]

| **g DM/day/kg $W^{0.75}$** | Sheep (S) | Goat (G) | S/G ratio |
|---|---|---|---|
| **Fresh state** | | | |
| *A. halimus(AH)* | 19.4 | 15.8 | 1.23 |
| *H. strobilaceum (HS)* | 18.2 | 14.2 | 1.28 |
| *T. aphylla (TA)* | 9.69 | 8.72 | 1.11 |
| *T. mannifera (TM)* | 10.9 | 10.3 | 1.06 |
| *Z. album (ZA)* | 2.12 | 2.94 | 0.72 |
| **Air- dried state** | | | |
| *A. halimus* | 13.2 | 21.0 | 0.63 |
| *H. strobilaceum* | 9.45 | 8.23 | 1.15 |
| *T. aphylla* | 0.00 | 0.00 | 0.00 |
| *T. mannifera* | 5.12 | 4.75 | 1.08 |
| *Z. album* | 0.00 | 0.00 | 0.00 |
| **Silage state** | | | |
| *AH+BL+MO** | 59.9 | 59.7 | 1.00 |
| *HS+BL+MO* | 58.2 | 59.4 | 0.98 |
| *TM+BL+MO* | 46.5 | 42.9 | 1.08 |
| *TA+BL+MO* | 44.2 | 42.2 | 1.05 |
| *ZA+BL+MO* | 24.2 | 22.5 | 1.09 |
| *Shrubs mixture** +GDS* | 37.7 | 35.8 | 1.05 |
| *Shrubs mixture+ GDS+Urea* | 42.4 | 30.9 | 1.37 |
| *Shrubs mixture +FFB* | 18.0 | 11.2 | 1.61 |
| *Shrubs mixture +FFB +Urea* | 20.8 | 24.1 | 0.86 |

*BL: broiler litters; MO: molasses; GDS: ground date seeds; FFB: fresh fodder beet ** Shrubs mixture: AH+ HS+ TM+ ZA at 1:1:1:1

## Utilization of halophytes by animal species

### *Small ruminants*

Sheep and goats, as the most dominant animal species in arid regions, could be efficiently fed on halophytic forages as the main source of feeding materials. These animals should be supplemented with any available energy non-saline feed resources to enhance the utilization of halophyte feed materials. Crushed or ground barley grains, ground date seeds and molasses are commonly used as energy supplements for sheep and goats fed salt marsh plants. It was noticed that halophytes intake was affected by the levels of energy supplement; animals given barley grains showed higher forage consumption than those fed saltbush as a sole diet (Tab. 5). Barley grains improved the growth rate of sheep and goats fed on *Atriplex* spp and decreased water intake [8]. The daily amounts of 150 and 250 g barley grains/head were recommended for sheep fed on saltbush during all seasons [16, 23]. The effect of energy supplements on pregnant and lactating sheep and goats grazing the natural salt marsh plants in Sinai, Egypt were studied [6]. The grazing animals were supplemented with different levels of barley grains during two successive years. It appeared that both animal species could not sustain themselves on the natural ranges without supplements. Barely supplements improved the DM and nitrogen intakes of the ranges. The highest values were recorded for both sheep and goats given the supplement (100 % of the maintenance requirements). The performance of grazing dams during the pregnancy and lactation stages, in terms of body weight changes and milk yield, in addition to daily gain and weaning weight of their offspring were improved by the energy supplementation. Therefore, it was recommended to give additional supplements, particularly energy source, for productive animals grazing on halophytic pastures.

The effect of feeding six silages made from combinations of natural and cultivated halophytic shrubs plus a mixture of fresh cultivated shrubs of *Atiplex nummularia* and *Aacacia saligna* on the performance of six groups of sheep was investigated [22]. The results of body weight changes of these sheep varied significantly ($P < 0.01$)

**Table 5.** Average values of saltbush intake during the wet season

| g/day/Kg $W^{0.75}$ | Sheep | Goat | Sources |
|---|---|---|---|
| *A. nummularia:* | | | |
| Solely fresh | 40.0 | – | [23] |
| + 150 g barley | 59.8 | 58.0 | [8] |
| + 250 g barley | 63.2 | 60.0 | [16] |
| *A. halimus* : | | | |
| Solely fresh | 19.4 | 15.8 | [29] |
| Solely fresh | 13.4 | 14.8 | [16] |
| Solely dry | 13.1 | 21.0 | [29] |
| + 100 g barley | 26.4 | 27.6 | [16] |
| + 150 g barley | 41.1 | 38.2 | [16] |

among treatments. Although the sheep used in this study were mature, some of them gained noticeable weight whereas others lost weight due to the fluctuations of the voluntary feed intakes pattern.

*Camels*

Camels are well adapted to arid and semi-arid regions, particularly to desert areas, where other animal species do not thrive and perhaps do not survive [4]. They have the capacity to utilize low quality feed resources as halophytes and convert them into animal protein and other products [24]. Camels could maintain themselves on natural ranges based mainly on halophytic plant species [4]. Some studies have been carried out on camels fed some halophytes in Egypt. 16 adult female camels were used to study the effect of energy level supplementation on the intake and utilization of *Atriplex nummularia* supplemented with yellow corn grain to cover none (group A or control), 20 % (group B), 40 % (group C) and 60 % (group D) of maintenance requirements of energy [24]. It is indicated (Tab. 6) that the voluntary intake of saltbush (*A. nummularia*) was significantly ($P < 0.01$) increased as a result of increasing the energy supplementation level which in turn was reflected on improving the body weight gain. The nutrients digestibility and nutritive values were also improved up to a level of 40 %. Studies on fattening camel on saltbush (*Atriplex halimus*) with different energy sources was conducted to evaluate growth performance, efficiency of feed utilization and carcass traits of male calves [24]. Daily intakes from DM, TDN and DCP were not affected by the experimental diets (Tab. 6). At the same

**Table 6.** Feed utilization of camel calves fed on various diets based on *Atriplex halimus* [24]

| Item | A | B | C | SEM |
|---|---|---|---|---|
| Initial live weight (kg) | 177.0 | 176.0 | 180.0 | 5.32 |
| Daily gain (g/day) | 750.0 | 732.2 | 692.2 | 0.03 |
| Daily feed intake (kg/head): | | | | |
| Concentrate mixture | 2.68 | – | – | – |
| Barley grains | – | 2.47 | 1.87 | – |
| Olive cake | – | – | 0.85 | – |
| Berseem hay | 4.00 | – | – | – |
| Fresh saltbush | – | 13.00 | 11.00 | – |
| Daily nutrient intake (kg): | | | | |
| Kg DM/head | 6.02 | 5.37 | 5.06 | 1.05 |
| Kg TDN/head | 3.75 | 3.36 | 3.65 | 0.69 |
| Feed conversion: | | | | |
| Kg DM/head | 8.03 | 7.33 | 7.31 | 0.63 |
| Feed cost /kg gain (US$) | 0.76 | 0.38 | 0.36 | – |

Diet A = conventional concentrate + berseem hay (control diet), Diet B = ground barley + *Atriplex hamus* (fresh saltbush), Diet C = 75 % ground barley + 25 % olive cake + *Atriplex halimus* (fresh saltbush)

time, calves fed on saltbush and barley grains either only or with olive improved feed conversion in terms of kg TDN/Kg weight gain and Kg DCP/kg gain. Feeding camel calves on fresh saltbush resulted in appreciable reduction in feeding cost for production of one Kg body weight. Feeding camel calves on diets B and C reduced the cost of feeding by about 50 % and 53 %, respectively, compared to the control diet. It is clearly shown that fresh saltbush can be successfully and economically used in feeding camel calves in arid and semi arid zones.

*Poultry*

There is a shortage of feedstuffs used in poultry feeding in most of arid countries. Therefore, using non-traditional ingredients in poultry feeding can substantially contribute to solving this problem by decreasing the feeding cost. Research studies conducted in Egypt showed that *Atriplex* and *Acacia* leaf meals have a reasonable content of crude protein (16.20–19.95 %). Table 7 summarizes data on some nutrients and minerals in *Atriplex* leaf meal derived by some authors at the Desert Research Center, in Egypt. Data on *Acacia* leaf meal (ACLM) also reported a higher value of crude fiber, while ash content revealed a pronounced higher value in *Atripex* leaf meal (ATLM) as illustrated in [25]. The ATLM and ACLM could be successfully used as a non conventional feedstuff for poultry species as shown in Table 8. In this respect, [26] used ATLM safely in mash form at a level of 6 % and 12 % for growing rabbits and layer turkeys, respectively. Pelleted ACLM and ATLM were used for growing rabbits with up to 20 % and 25 % of the total diets [25, 27]. No effects on body weight gain, feed utilization and carcass characteristics were observed.

The impact of using such halophytic feed materials in poultry feeding greatly vary according to species, nutrient content of diet and some other environmental factors. Therefore, the recommended levels of ACLM and ATLM in poultry diets varied greatly. For example, ATLM was recommended at a level of 25 % in rabbit diets. This value was reduced to 6–12 % in growing layer turkeys. ACLM ranged from

**Table 7.** Chemical composition of *Atriplex* leaf meal (ATLM), on DM basis

| **Items** | **Source[26]** | **Source[27]** |
|---|---|---|
| Crude Protein, % | 19.95 | 18.89 |
| Crude Fiber, % | 2.43 | 6.26 |
| Ether Extract, % | 3.06 | 3.11 |
| Nitrogen-free extract, % | 40.91 | 41.33 |
| Ash, % | 22.05 | 20.19 |
| Calcium, % | 1.40 | 1.99 |
| Phosphorus, % | 0.40 | 0.24 |
| Potassium, % | 3.20 | 4.82 |
| Sodium, % | 5.90 | 8.96 |
| Iron, ppm | 0 | 21.0 |
| Copper, ppm | 78.0 | 28.0 |
| Zinc , ppm | 46.0 | 0 |

**Table 8.** Recommended levels of ATLM and ACLM percentages in poultry diets

| Species | Author | ATLM | ACLM | Type of feeding |
|---|---|---|---|---|
| Growing rabbits | [30] | – | 40 | Wilting green leaves |
| Growing rabbits | [28] | – | 15 | Pellet |
| Growing rabbits | [25] | | 20 | Pellet |
| Growing rabbits | [27] | 25 | – | Pellet |
| Growing rabbits | [26] | 6 | – | Mash |
| Layer turkeys | [26] | 12 | – | Mash |
| Broilers | [26] | – | 6 | Mash |

15–30 % in a pelleted form for growing rabbits without detrimental effects on body weight gain and feed utilization, whereas this value decreased to 6 % for broiler diets. Increasing ACLM and ATLM in poultry diets was followed by an increase in total and in daily feed consumption. However, higher levels of ACLM and ATLM above the recommended values coincided with the decrease in body weight (daily gain and feed conversion). At the same time, nutritive value decreased due to the decreasing digestion coefficients of its nutrient content. Different levels of ACLM and ATLM had no significant effect on carcass characteristics and chemical components of meat in poultry. It is noted that increasing ACLM and ATLM in growing rabbit's diets resulted in an increased total and daily feed consumption [28, 25, 27]. This indicated that the inclusion of such ingredients in diets might increase their palatability. In this concern, [26] showed that inclusion of ATLM in layer and growing turkey's diets up to 6 % and 12 %, respectively, increased significantly the feed intake of birds as compared to that of the control group. The inclusion of ACLM and ATLM at higher levels caused a higher percentage of mortality (15 % at levels of 40 % ACLM and 30 % ATLM, respectively). This might be due to presence of some anti-nutritional factors in such halophytes (e.g., tannin, oxalate, etc.).

The lower costs of ACLM and ATLM are reflected on the feeding cost, net return and the values of economical efficiency. However, although ACLM and ATLM are non-traditional feedstuffs that could be used in poultry diets, attention must be given to their chemical composition, particularly their content of the anti-nutritional factors.

## Conclusion

Halophytes for utilization are available in various regions and under different climates. The benefits of such use for the local freshwater saving are obvious, particularly in arid regions. Some applications of halophytes are highly profitable, for production of additional human food and for animal fodder. It is imperative that commercial production of halophytes should now be taken from research to large-scale field trials. Prospects of both short- and long-term production trials are well recognized and established. Halophytes could be a traditional source of animal nutrition even though they may present some problems.

## References


1 Choukr-Allah R, Malcom CV, Hamdy A (1996) *Halophytes and biosaline agriculture.* Marcel Dekker, New York, 400

2 Lieth H, Moschenko M, Lohman M, Koyoro HW, Hamdy A (1999) Halophyte uses in different climates. 1. *Ecological and ecophysiological studies.* Progress in Biotechnology, Vol. 13: 258. Leiden, Backhuys Publishers

3 Squires VR, Ayoub AT (1994) *Halophytes as a resource for livestock and for rehabilitation of degraded lands.* Kluwer Academic Publishers, The Netherlands, 123

4 Wilson RT (1984) *The Camel.* Longman, London and New York

5 Le Houerou HN (1994) Forage halophytes and salt tolerance fodder crops in the Mediterranean Basin. In: VR Squires, AT Ayoub (eds) In: *Halophytes as a resource for livestock and for rehabilitation of degraded lands.* Kluwer Academic Publishers, The Netherlands, 123

6 El Shaer HM (1981) A comparative nutrition study on sheep and goats grazing Southern Sinai desert range with supplements. Ph.D. Thesis, Fac Agric Ain Shams Univ Egypt

7 Arieli A, Naim E, Benjamin RW, Pasternak D (1989) The effect of feeding saltbush and sodium chloride on energy metabolism in sheep. *Anim Prod* 49: 451–457

8 Kandil HM, El Shaer HM (1988) The utilization of *Atriplex nummularia* by goats and sheep in Sinai. Proc Inter Symp on the Constraints and Possibilities of Ruminant Production in Dry Subtropics, 5–7 November 1988. Cairo, Egypt

9 El Shaer HM (1997) Practical approaches for improving utilization of feed resources under extensive production system in Sinai. Proc Inter Symp on Systems of Sheep and Goat Production, 25–27 October 1997. Bella, Italy

10 Atiq-ur-Rehman (2002) Utilization of *Atriplex* as a forage under grazing and cut and carry systems, for small ruminants. Proc Intr Symp on optimum resources utilization in salt-affected ecosystems in arid and semi-arid regions, 8–11 April 2002. Cairo, Egypt

11 Koocheki A, Mohalati M (1994) Feed value of some halophytic range plants of arid regions of Iran. In: VR Squires, AT Ayoub (eds) In: *Halophytes as a resource for livestock and for rehabilitation of degraded lands.* Kluwer Academic Publishers, The Netherlands, 249–253

12 El Shaer HM (1996) Rangelands as feed resources in the Egyptian desert: Management and improvement. Proc of the Inter Conf on Desert Development in the Arab Gulf countries, 23–26 March 1996. State of Kuwait

13 Gihad EA, El Shaer HM (1994) Nutritive value of halophytes. In: VR Squires, AT Ayoub (eds) In: *Halophytes as a resource for livestock and for rehabilitation of degraded lands.* Kluwer Academic Publishers, The Netherlands, 281–284

14 Le Houerou HN (1993) Salt tolerance plants for the arid region of the Mediterranean Isoclimatic zone. In: H Leith, A El-Masoom (eds) In: *Towards the rational use of high salinity tolerant plants.* Vol 1: 403. Kluwer Academic Publishers, Dordrecht, The Netherlands

15 Warren BE, Casson T (1992) Saltbush quality and sheep performance. Proc of a Workshop on Productive use of saline land, 10–14 May 1991. Perth, Western Australia 71–74

16 El Shaer HM, Kandil HM (1990) Comparative study on the nutritional value of wild and cultivated *Atriplex halimus* by sheep and goat in Sinai. *Com Sci Dev Res* 29: 81–90

17 Nawaz S, Hanjra SH (1993) The productivity of goats using different ratio of salt bush and kallar grass. In: NJ Davidson, R Galloway (eds) In: *Productive use of saline land.* ACIAR Proc No 42: 85–87. Canberra, Australia

18 El Shaer HM, Kandil HM, Khamis HS (1991) Salt marsh plants ensiled with dried broiler litter as feedstuff for sheep and goats. *Agric Sci Mansoura Univ* 16: 1524

19 Abd El Rahman HM (1996) Utilization of desert range poor quality feeds by sheep and goats. M.Sc. Thesis, Fac Agric, Cairo University, Egypt

20 Nasr SN (2001) Alteration of blood components and histopathological changes in goats exposed to some range plants growing in the Egyptian desert. Ph.D. Thesis, Fac of Vet Medic, Cairo University, Cairo, Egypt
21 Abou El-Nasr HM, Kandil HM, El-Kerdawy DA, Khamis HS, El-Shaer HM (1996) Value of processed saltbush and Acacia shrubs as sheep fodder under arid conditions of Egypt. *Small Ruminants Res* 24: 15–20
22 Fahmy A, El-Shaer HM, Shalaby AS, Abdul Aziz GM, Abd El-Gawad M (2002) Palatability and nutritional quality of some halophytic silages fed to sheep in Sinai. Proc Intr Symp on optimum resources utilization in salt-affected ecosystems in arid and semi-arid regions, 8–11 April 2002, Cairo, Egypt
23 Hassan NI, Abd El Aziz HM (1979) Effect of barley supplementation on the nutritive value of saltbush (*Atriplex nummularia*).*World Rev Anim Prod* 15: 47–55
24 Shawket Safinaz M (1999) Fattening of camels calves on saltbush (*Atriplex halimus*) with different energy sources. *J Agric Sci Mansoura Univ* 24: 1751–1764
25 Abd El Galil K, Khider RE (2000) Utilization of *Acacia saligna* in feeding growing rabbits under the desert and newly reclaimed areas. *Egypt Polut Sci J* 20: 497–515
26 Amin E (1999) Effects of cross breeding and feeding different levels of *Atriplex* meal on production characters of local and commercial verities of turkey. M.Sc. Thesis, Fac of Agric Alexandria University, Alexandria, Egypt
27 Abd El Galil K, Khider RE (2001) Utilization of *Atriplex nummularia* in feeding growing rabbits under the desert and newly reclaimed areas. *Egypt Polut Sci J* 21: 53–71
28 El-Eraky WL, Mohamed WM (1996) Growth performance, carcass traits and some related physiological changes of growing rabbits fed on acacia and water hyacinth. *Egypt J of Rabbit Sci* 6: 87–98
29 El Shaer HM, Salem OA, Khamis HS, Shalaby AS, Farid MFA (1990) Nutritional comparison studies on goats and sheep fed broiler litter ensiled with desert shrubs in Sinai. Proc of the Inter Goat Production Symp, 22–26 October 1990. Tallahassee, Fl, USA, 70–74
30 Abd el Samee AM, El-Gendy KM, Ibrahim H (1992) Growth performance and some related physiological changes in rabbits as affected by feeding Acacia under sub-tropical conditions. *Egyptian J Rabbits Sci* 2: 13–22

Biosaline Agriculture and Salinity Tolerance in Plants
Edited by M. Öztürk, Y. Waisel, M.A. Khan and G. Görk


# Potential of dry drainage as a sustainable solution to waterlogging and salinisation

Fatih Konukcu

*Trakya University, Tekirdag Agricultural Faculty, Irrigation and Drainage Dept., TR-59030 Tekirdag, Turkey*

## Introduction

It is estimated that, of the total 270 million hectares of irrigated land around the world, some 80 million hectares are affected by waterlogging and salinity, while about 20 million hectares suffer from severe irrigation-induced salinity problems [1].

Conventional wisdom holds that the best solution to dealing with salinity and waterlogging is to maintain a net flux of salt away from the rootzone and to control the water table by artificial drainage. A particular drainage may be suitable for local circumstances, but with large contiguous irrigation systems, there are two important general limitations, economic and environmental [2].

In recent years, there have been attempts to identify solutions which will work within environmental constraints and will also be economically viable [2, 3]. Improved on-farm water management combined with disposal by means of evaporation ponds is seen as the reasonable strategy, though with some environmental risks [3]. Another alternative is the control of the water level with irrigation management. A shallow water table can be considered as a valuable resource for meeting part of the crop requirement for water [4, 5]. However, in arid and semi-arid regions, the evaporative demand and the salinity of groundwater are usually high and the upward evaporative flux from a shallow saline water table results in the accumulation of salt to a very high concentration at or near the soil surface.

Salinity control depends upon establishment of a time averaged net downward flux through the rootzone. Therefore, it is the water balance that is important in determination of the balance. Within a given area, if inflow balances outflow, then the water table will be stable. If the non-cropped area is large enough and evaporation from this area is fast enough, then the necessary balance can be achieved without artificial drainage. This is the concept of dry-drainage. The groundwater system provides the pathway for the movement of the excess water from the irrigated land to the fallow land (Fig. 1).

The objectives of this study are i) to describe water and salt balances of both cropped and fallow areas, and ii) to assess the requirements for a successful dry-drainage system.

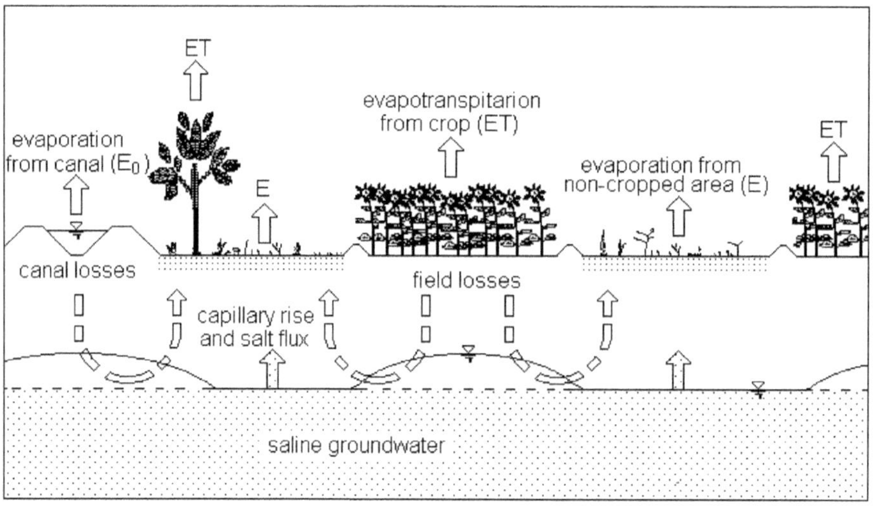


**Figure 1.** Schematic section of a dry-drainage system

## Theory

As shown in Figure 1, a single dry-drainage unit has two parts, the irrigated area and the fallow area.

### *Design of the irrigated field*

In arid and semi-arid areas, evapotranspiration (ET) exceeds the precipitation (P) and the resulting water deficit should be covered by irrigation (I) to achieve a satisfactory yield. The application of irrigation water means an input of salts because irrigation water, even if of excellent quality, is a major source of soluble salts. If soil salinisation is to be avoided, these salts have to be leached out by deliberate over-irrigation with the solution percolating to the subsoil. The leaching requirement ($R^x$) to provide the salt balance of the rootzone may be calculated following van Hoorn and van Alphen [6]. The total irrigation (I) then includes the leaching requirement as well.

The irrigation schedule is designed to maintain the salt balance during the whole year. Salt accumulation and leaching requirement of the irrigated area during the fallow period is computed as is in the fallow field.

### *Design of the fallow area*

In investigating the sustainability of a dry-drainage system, we need to predict accurately the rate of evaporation and salt accumulation in the fallow area. Gowing et al. [7] developed a pseudo steady-state model, modifying the well-known Gardner [8] model, to predict the rate of evaporation from the soil surface, particularly from the surface of bare soil. Therefore, this model will be used to compute the rate of evaporation from the fallow area. Accordingly they locate the depth of Evaporation Front (EF) and then calculate the rate of evaporation. The investigators have distinguished three stages in the progression of the EF: i) no EF exists, ii) the EF moves downwards, and iii) the EF is stationary [9]. The governing equations and calculation procedures for each of the stages are fully described in [7].

The concentration profile which develops with time depends upon both the upward evaporative flux of water, which concentrates salts, at the surface, and the diffusive–dispersive flux, which tends to move salts downward against the upward flux of water. Elrick et al. [10] model is applied to simulate salinity profile of the fallow area.

In the leaching process, the soil profile is considered as a series of one-dimensional reservoirs with bypass both in the fallow area after an effective rainfall and in the cropped area after a fallow period [6].

*Parameters for simulating field behaviour*

Average climatologic data and soil properties for the Lower Indus Basin in Pakistan are adopted from Asghar [2] (Tab. 1) and estimates of evapotranspiration (ET), are from Gowing and Wyseure [3] (Tab. 2). Salinity of the irrigation water, $C_i$, is 0.7 g/l [2, 3] whereas the salinity of the drainage water, $C_s$, is 2.8 g/l for a given crop pattern. The groundwater salinity of 7.0 g/l is taken in this study throughout the simulation period. A sandy clay loam soil prevails in the region [2]. The water table depth is taken as 1.5 m [2]. Monthly average precipitation (P) is distributed within a given month over an equal period taking the number of rains into consideration.

**Table 1.** Average climatologic data in the study area [2]. $T$, mean daily temperature; $h_a$, mean relative humidity of air; $P$, monthly mean precipitation and $N$, number of rains

| Months | J | F | M | A | M | J | J | A | S | O | N | D |
|---|---|---|---|---|---|---|---|---|---|---|---|---|
| $T$(° C) | 13 | 16.7 | 22.5 | 28.1 | 33.3 | 33.6 | 31.4 | 30.3 | 28.9 | 26.1 | 20.0 | 15.3 |
| $h_a$ (%) | 57 | 51 | 36 | 27 | 28 | 45 | 67 | 72 | 62 | 44 | 41 | 56 |
| $P$(mm) | 23 | 18 | 13 | 8 | 13 | 74 | 180 | 173 | 117 | 10 | 3 | 10 |
| N (d) | 2 | 2 | 1 | 1 | 2 | 4 | 8 | 8 | 4 | 1 | 1 | 1 |

**Table 2.** Average evapotranspiration [3], irrigation and leaching requirements in the study area. $ET_0$, reference evapotranspiration; $ET$, evapotranspiration for wheat and cotton; $I^\times$, irrigation without leaching ($I^\times = ET - P$); $I$, total irrigation amount ($I = I^\times + R^\times$) and $R^\times$, leaching requirement with 80 % leaching efficiency coefficient $f_i$+ field losses of 15 %)

| Month | J | F | M | A | M | J | J | A | S | O | N | D |
|---|---|---|---|---|---|---|---|---|---|---|---|---|
| $ET_0$ (mm) | 64 | 82 | 140 | 183 | 243 | 262 | 214 | 198 | 177 | 136 | 81 | 61 |
| $ET$ wheat (mm) | 60 | 90 | 98 | | | | | | | | 75 | 34 |
| $ET$ cotton (mm) | | | | 75 | 49 | 144 | 225 | 218 | 177 | 109 | | |
| $I^\times$(mm) | 59 | 96 | 113 | 89 | 48 | 93 | 60 | 60 | 80 | 132 | 96 | 32 |
| $R^\times$ (mm) | 10 | 17 | 20 | 16 | 8 | 16 | 11 | 11 | 14 | 23 | 17 | 16 |
| $I$(mm) | 69 | 113 | 133 | 105 | 56 | 109 | 71 | 71 | 94 | 155 | 113 | 48 |

## Results and discussion

*Water and salt balance of the cropped area*

The amount of water (leaching + irrigation losses) percolating from the cropped area for each month during a year is given in Table 2. The maximum and minimum percolation occurred in October (27 mm) and December (7 mm), respectively.

*Water and salt balance of the fallow area*

Figure 2a shows the daily evaporation from the soil surface over a year: note that the calculation started from October but is presented from January. A relatively high evaporation rate on the first day (day 273), about 8 mm/d, decreased to 2.1 mm/d within the first 7 days because the evaporative demand of the atmosphere exceeded the ability of the soil to conduct water so causing the soil surface to dry. The evaporation rate then fluctuated minimally above this value during the dry season following small amounts of precipitation. Daily evaporation increased suddenly when the rainy season began and then fluctuated widely between the potential and limiting rates during the rainy season.

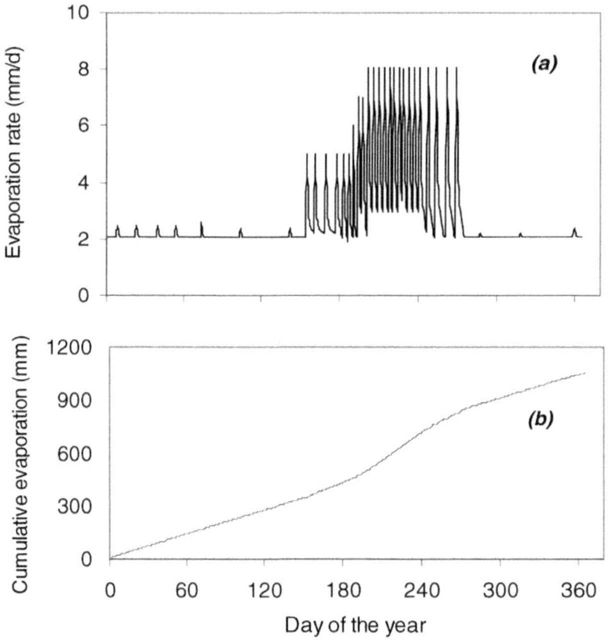


**Figure 2.** (**a**) Evaporation rate and (**b**) cumulative evaporation from the soil surface of the fallow area during a year

The cumulative evaporation from the fallow area should balance the total of precipitation and percolating water from the cropped area for dry-drainage to be a success. The cumulative evaporation from the fallow area was 1,054 mm/year (Fig. 2b) while the sum of precipitation (643 mm/year) and percolating water from the cropped area (198 mm/year) amounted to 841 mm/year. This means that the fallow field is capable of sustaining the required water balance for the success of the system. Under the simulated conditions, the cropped area may be larger than the fallow area by a factor of 1.25 (i.e., 1,054/841).

The salt-concentration profile was calculated monthly. The salt and water profiles at the end of the previous month were used as the initial conditions for the next month.

Figure 3 shows the calculated salt concentration profiles at four different times during the dry period. At the end of the dry season, approximately the top 60 cm of soil layer had become saline. Salt concentration in the top 10 and 20 cm did not change after the end of January and May because of the formation of an evaporation front in which water was considered to move as vapour. Leaching was calculated during the rainy season.

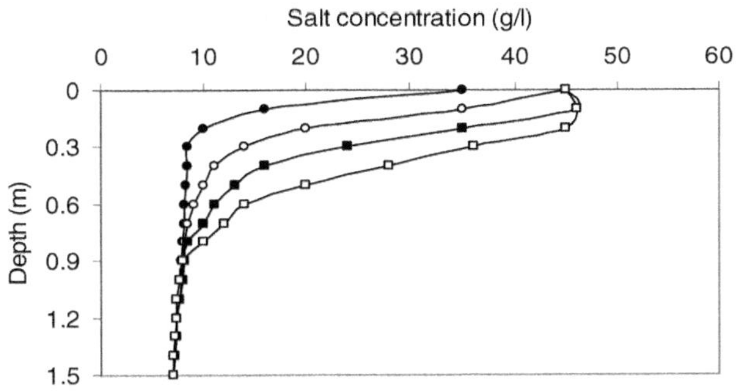


**Figure 3.** Profiles of salt concentration of the fallow area at four different times during the dry period. •: 30 November, ○: 31 January, ■: 31 March, □: 31 May

After each rain, the water and salinity profiles were recalculated. Figure 4 shows the calculated salt concentration of each soil layer. The amount of evaporation during the period between two rainfalls was allowed for calculating the next water-content profile. Although the rainy season started from June, the amount of precipitation during this month was not sufficient to replenish the water content to field capacity so no percolation and therefore no leaching occurred. At the end of July, a considerable amount of salt from the 0–30 cm soil layer was leached into the 30–60 cm layer but there was no leaching below the 30–60 cm layer. During August and September, leaching occurred in all soil layers; however, a small amount of salt still remained in the soil at the end of the year.

Having carefully considered the water and salt balance of both irrigated and fallow area, the remaining salt in the soil profile of the fallow area at the end of the year may be leached if this leaching requirement is not too large, as practiced

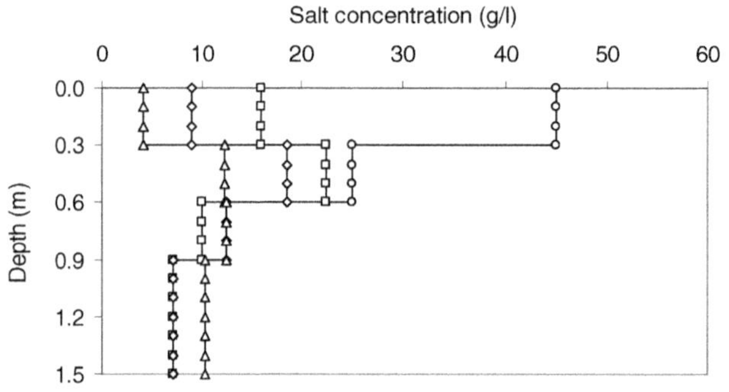


**Figure 4.** Changes in the salt concentration of the soil layers leached after four different times during the rainy period. ○: initial, □: 30 July, ◇: 31 August, △: 30 September

in West Africa [2]. In our case, 120 mm water is needed. Re-checking the water balance of the fallow land, the amount of inflow (961 mm) needed, which is the sum of percolating water from the irrigated area (198 mm), total precipitation (643 mm) and leaching requirement of fallow field (120 mm), is still smaller than that of the outflow which is the cumulative evaporation from the fallow field, 1,054 mm. In this case, the irrigated area may be approximately 10 % larger than the fallow area, which is virtually the same as proposed by [2, 3].

## Conclusions

The rate of evaporation the fallow area was fast enough to cover the net downward flux through the cropped area due to the leaching requirement and 15 % irrigation losses when the groundwater was stable at 1.5 m depth. Under the simulated conditions, the necessary water balance could be achieved when the cropped area was larger than the fallow area by a factor of 1.25. However, the salt balance of the fallow area was not satisfied. If this salt is to be leached at the end of the growing season, the ratio of cropped to fallow areas becomes 1:10. However, this ratio may change with climate, soil type, water table depth, irrigation amount and groundwater quality and crop pattern.

About 50 % of the potentially irrigable land should be abandoned, which may be hard to accept. But, given the choice of paying the full economic cost of a regional drainage scheme, would farmers find it so unattractive? A full economic analysis is needed.

## References

1 Kandiah A (1990) Water soil and crop management. Relating to the use of saline water. FAO Expert Consultation, AGI2MISC/16/90, 193
2 Asghar MN (1996) Computer simulation of salinity control by means of an evaporative sink. PhD thesis. University of Newcastle upon Tyne, UK

3 Gowing JW, Wyseure GCL (1992) Dry-drainage a sustainable and cost-effective solution to waterlogging and salinisation. Proceedings of $5^{th}$ International Drainage Workshop, ICID-CIID, Lahore, Pakistan, Vol 3: 6.26–6.34
4 Ragab RA, Amer F (1986) Estimating water table contribution to the water supply of maize. *Agriculture Water Management* 11: 221–230
5 Qadir M, Oster D (2004) Crop and irrigation management strategies for saline-sodic soils and waters aimed at environmentally sustainable agriculture. *Science of the Total Environment* 323(1–39): 1–19
6 van Hoorn JW, van Alphen JG (1994) Salinity control. In: HP Ritzema (ed.) In: *Drainage Principles and Applications*. International Institute for Land Reclamation and Improvement, Wageningen, The Netherlands, 533-600
7 Gowing JW, Konukcu F, Rose DA (2006) Evaporative flux from a shallow water table: the influence of a vapour–liquid phase transition. *J Hydrology* (in press)
8 Gardner WR (1958) Some steady-state solutions of the unsaturated moisture flow equation with applications to evaporation from a watertable. *Soil Science* 85: 228–232
9 Rose DA, Konukcu F, Gowing JW (2005) Effect of watertable depth on evaporation and salt accumulation above saline groundwater. *Australian J Soil Res* 43: 565–573
10 Elrick DA, Mermoud A, Monnier T (1994) An analysis of solute accumulation during steady-state evaporation in an initially contaminated soil. *J Hydrology* 155: 27–38

Biosaline Agriculture and Salinity Tolerance in Plants
Edited by M. Öztürk, Y. Waisel, M.A. Khan and G. Görk


# Vegetative bioremediation of sodic and saline-sodic soils for productivity enhancement and environment conservation

Manzoor Qadir[1], Jim D. Oster[2], Sven Schubert[3] and Ghulam Murtaza[4]

[1] *International Center for Agricultural Research in the Dry Areas, PO Box 5466, Aleppo, Syria*
[2] *Department of Environmental Sciences, University of California, USA*
[3] *Institute of Plant Nutrition, Justus Liebig University, Giessen, Germany*
[4] *Institute of Soil and Environmental Sciences, University of Agriculture, Faisalabad, Pakistan*

## Introduction

Salt-affected soils occupy nearly 20 % of irrigated area worldwide [1]. As a major category of salt-affected soils, sodic and saline-sodic soils are characterized by the occurrence of sodium ($Na^+$) at levels that result in poor physical properties and fertility problems, thereby threatening agricultural productivity in many arid and semi-arid regions. Amelioration of these soils is driven by providing a soluble source of calcium ($Ca^{2+}$) to replace excess $Na^+$ on the cation exchange complex [2]. The displaced $Na^+$ is either leached from the root zone by excess irrigation, a process that requires soil permeability and provision of a natural or artificial drainage system, or is taken up by crops.

Many sodic and saline-sodic soils contain inherent or precipitated sources of $Ca^{2+}$, i.e., calcite ($CaCO_3$) at varying depths. Due to its negligible solubility (0.14 mmol $L^{-1}$), natural dissolution of calcite does not provide sufficient $Ca^{2+}$ to ameliorate these soils. Consequently, amelioration of these soils has been dominated by the application of chemical amendments [3]. Some amendments supply soluble sources of $Ca^{2+}$ to the soil solution, which then replace excess $Na^+$ on the exchange complex, while others assist in increasing the dissolution rate of calcite. There have been constraints with chemical amelioration in several developing countries during the last two decades because of 1) low-quality of amendments containing a large fraction of impurities, 2) restricted availability of amendments when needed for amelioration, and/or 3) increased costs of amendments due to competing demands by industry and reductions in government subsidies for their agricultural use. Owing to overriding importance of the last factor, chemical amelioration has become prohibitively expensive for resource-poor farmers. However, there is growing evidence from researchers and farmers indicating that these soils can be brought back to a highly productive state using a plant-assisted amelioration approach – vegetative bioremediation – that does not rely on chemical amendments [4–6]. Synonymous terms for vegetative bioremediation include phytomelioration, phytoremediation, and biological reclamation.

Typical plant-assisted amelioration strategies for contaminated soils, such as those containing elevated levels of heavy metals and metalloids, work through cultivation of specific plant species capable of hyper-accumulation of target ionic species in their shoots, thereby removing them from the soil [7]. In contrast, vegetative bioremediation of sodic and saline-sodic soils is achieved by the ability of plant roots to increase the dissolution rate of calcite, thereby resulting in enhanced levels of $Ca^{2+}$ in soil solution. The salinity-sodicity combination present in the soil solution during vegetative bioremediation maintains adequate soil structure and aggregate stability that enhance the amelioration process [8]. This chapter highlights the role of cropping for vegetative bioremediation of calcareous sodic and saline-sodic soils and its evaluation against other amelioration approaches. This information will assist researchers and farm advisors in choosing appropriate crops as well as crop, soil and irrigation management practices to achieve maximum benefit during the amelioration process.

## Vegetative bioremediation of sodic and saline-sodic soils

Vegetative bioremediation of calcareous sodic and saline-sodic soils is a promising option that increases the dissolution rate of calcite through the processes at the soil-root interface resulting in enhanced levels of $Ca^{2+}$ in soil solution. Vegetative bioremediation ($V_{\mathrm{Bio}}$) is a function of the following factors:

$$V_{\mathrm{Bio}} = \sum R_{\mathrm{PCO2}} + R_{\mathrm{H+}} + R_{\mathrm{Phy}} + S_{\mathrm{Na+}} \quad (1)$$

where $R_{\mathrm{P_{CO_2}}}$ refers to increased partial pressure of $CO_2$ within the root zone, $R_{\mathrm{H+}}$ is enhanced proton ($H^+$) release in the root zone in case of certain $N_2$-fixing crops, $R_{\mathrm{Phy}}$ deals with physical effects of roots in improving soil aggregation and hydraulic properties of the root zone, and $S_{\mathrm{Na+}}$ consists of $Na^+$ content of shoot which is removed through harvest of aerial plant portion. The collective effects of these factors ultimately lead to soil amelioration, provided leaching and drainage are adequate (Fig. 1).

## Comparative efficiency of vegetative bioremediation

The evaluation of vegetative bioremediation and chemical approaches in various countries reveals comparable performance of both in terms of sodic soil amelioration. Results of a field experiment conducted on a barren, calcareous, alkali soil ($pH_{1:2}$ = 10.6, $EC_{1:2}$ = 2.7 dS $m^{-1}$, ESP = 94) indicated that the amelioration efficiency of two grasses, Para grass (*Brachiaria mutica* (Forssk.) Stapf) and Karnal grass (*Leptochloa fusca* (L.) Kunth), was comparable with soil application of gypsum at 12.5 Mg $ha^{-1}$ [9]. The yield of first rice (*Oryza sativa* L.) crop in the gypsum treatment averaged 3.7 Mg $ha^{-1}$ as compared to 3.8 and 4.1 Mg $ha^{-1}$ from the treatments cropped for 1 year with Para and Karnal grasses, respectively. The corresponding

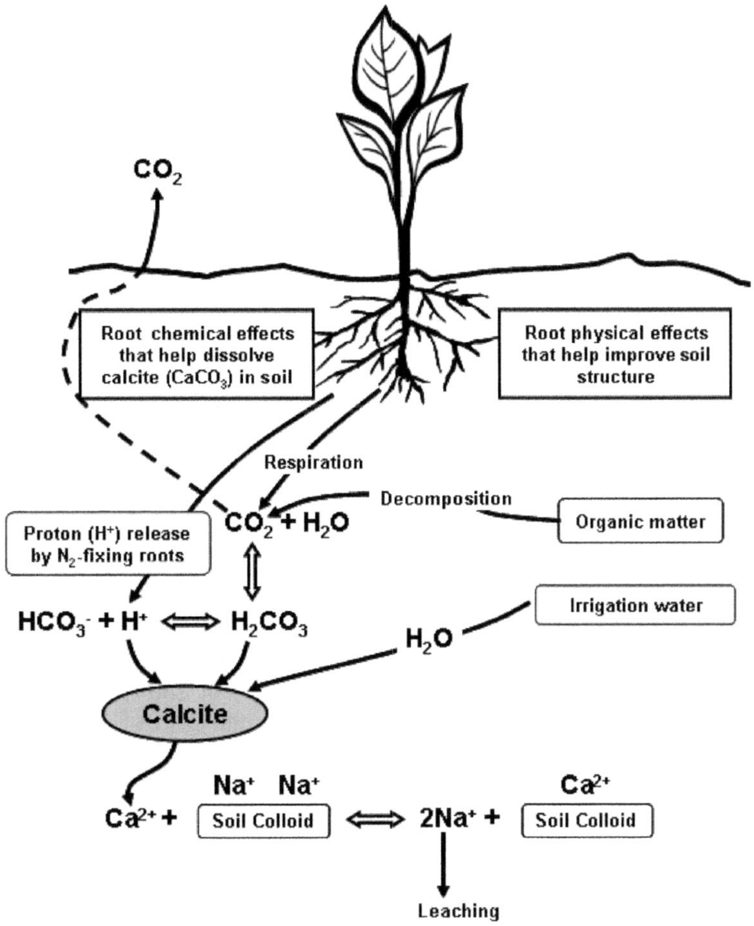


**Figure 1.** A conceptual model for the chemical reactions involved in calcite ($CaCO_3$) dissolution and amelioration of calcareous sodic and saline-sodic soils during vegetative bioremediation

rice yields after 2 years of grass cropping were 5.3 and 6.1 Mg ha$^{-1}$. In another field experiment [10], amelioration efficiency of Kallar grass was evaluated during different periods of root decay by leaching a calcareous, silty clay loam, saline-sodic soil (pH$_s$ = 8.3–9.3, EC$_e$ = 16.8–37.5 dS m$^{-1}$, SAR = 32.5–108.9) 3, 6, 9, and 12 days after each harvest during 2 years of grass cultivation. Each plot was kept flooded for 3 days during leaching. The amelioration efficiency of Kallar grass was greater in the plots leached 6 days after harvesting, and it was comparable with gypsum-treated soil.

In a field study [11], cropping of sesbania (*Sesbania bispinosa* (Linn.) W.F. Wight), Kallar grass, and sordan (*Sorghum* ×*drummondii* (Steud.) Millsp. & Chase) was compared against gypsum application (13 Mg ha$^{-1}$) on a calcareous, sandy clay loam, saline-sodic soil ($pH_s$ = 8.2–8.6, $EC_e$ = 7.4–9.0 dS $m^{-1}$, SAR = 55.6–73.0). The plant species were grown for two seasons (15 months) with average forage yields in the order: sesbania (40.8 Mg $ha^{-1}$) > Kallar grass (29.3 Mg $ha^{-1}$) > sordan (24.7 Mg $ha^{-1}$). After two cropping seasons, the treatment efficiency for grain yield of the subsequent wheat (*Triticum aestivum* L.) crop was in the order: sesbania (3.79 Mg $ha^{-1}$) ≈ gypsum (3.68 Mg $ha^{-1}$) > Kallar grass (3.14 Mg $ha^{-1}$) > sordan (2.27 Mg $ha^{-1}$) > control (0.65 Mg $ha^{-1}$). In a later field experiment [5], four plant species – Kallar grass, sesbania, millet rice, and finger millet – were tested against gypsum application (14.8 Mg $ha^{-1}$) to ameliorate a calcareous, sandy clay loam, saline-sodic soil ($EC_e$ = 9.1–11.0 dS $m^{-1}$, SAR = 59.4–72.4). The treatment effectiveness to decrease soil $EC_e$ and SAR was in the order: gypsum ≈ sesbania > Kallar grass > millet rice > finger millet. Forage yields of the plant species were directly proportional to their soil amelioration efficiency.

Some field trials of crop bioremediation techniques have not been successful primarily because a salt-resistant forage crop was not the first crop in the rotation. In a field experiment [12], biological (rice-wheat rotation), physical + biological (subsoiling by curved chisels to a depth of 0.5±0.05 m at a chisel spacing of 1.2–1.5 m + rotation), chemical + biological (gypsum at 100 % gypsum requirement of the upper 0.15 m of soil + rotation), and chemical + physical + biological (gypsum + subsoiling + rotation) methods were compared to ameliorate two calcareous saline-sodic soils. Irrigation water (EC = 1.8 dS $m^{-1}$, SAR = 9.8) was applied according to the crop water requirement. The first crop in the rotation was rice, which was a complete failure and did not produce any grain on one soil ($pH_s$ = 8.6–9.1, $EC_e$ = 12.3–15.0 dS $m^{-1}$, ESP = 58.7–74.6), and a grain yield of 0.72 Mg $ha^{-1}$ on the other soil ($pH_s$ = 8.8–8.9, $EC_e$ = 9.6–15.2 dS $m^{-1}$, ESP = 42.5–45.6). Four years after cropping, the average rice grain yield from both soils was in the order: gypsum (1.99 Mg $ha^{-1}$) > gypsum + subsoiling (1.84 Mg $ha^{-1}$) > subsoiling (1.41 Mg $ha^{-1}$) > vegetative bioremediation (1.02 Mg $ha^{-1}$). Gypsum and gypsum + subsoiling treatments had similar values for the wheat grain yield (2.72 Mg $ha^{-1}$) followed by subsoiling (1.79 Mg $ha^{-1}$) and vegetative bioremediation (1.46 Mg $ha^{-1}$). Within the upper 0.15 m depth, all the treatments decreased $EC_e$ levels less than 5 dS $m^{-1}$ and ESP levels less than 22 on both the soils.

Several crop rotations have been evaluated for the amelioration of saline-sodic and sodic soils. Three irrigated crop rotations were tested to ameliorate a calcareous saline-sodic field ($pH_s$ = 8.1–8.2, $EC_e$ = 9.2–13.7 dS $m^{-1}$, SAR = 30.6–42.7). The rotations distributed in plots of 18 $m^2$ were: sesbania-barley (*Hordeum vulgare* L.), rice-wheat, and Kallar grass-alfalfa (*Medicago sativa* L.). All the crop rotations reclaimed the upper 0.15 m of soil after 1 year (SAR < 10) as did amelioration by the non-cropped gypsum treatment, which decreased SAR less than 14 [13]. Although initial salinity and sodicity levels of this soil were closer to that used by [12], there were three differences: 1) the soil was relatively coarser in texture, 2) the plots were irrigated with canal water (EC = 0.3 dS $m^{-1}$, SAR = 0.5), and 3) the irrigation water was applied in excess of crop water needs to leach $Na^+$ to lower depths.

In an evaluation of 14 experiments, carried out in different parts of the world, there was a comparable effect of chemical and bioremediation approaches [14]. The chemical treatment (application of gypsum in all experiments) caused 62 % decrease in original sodicity levels, whereas a 52 % decrease was calculated for the vegetative bioremediation treatments. However, in some experiments bioremediation was either unsuccessful or much less efficient than the chemical treatment for the reasons: 1) a crop resistant to ambient soil salinity and sodicity levels was not the first in the crop rotation; 2) bioremediation crop was grown during the time, which was not its most suitable growing season; 3) time was insufficient to exploit the potential impact of the bioremediation crop; and/or 4) irrigation was not applied in excess of crop water requirement, which restricted the downward movement of $Na^+$ from the root zone. In general, bioremediation worked well on moderately sodic and saline-sodic soils, provided: 1) irrigation was in excess of crop water requirement to provide adequate leaching; and 2) the excess irrigation was applied when the crop growth and hence $P_{CO_2}$ were at their peak. On these soils, the performance of vegetative bioremediation was comparable with soil application of gypsum. On highly sodic and saline-sodic soils, chemical treatment was better than the cropped treatments.

## Additional benefits of vegetative bioremediation

Nutrient availability status of post-amelioration soil is crucial for the growth of subsequent crops. Research on nutrient behavior during amelioration using chemical and biological methods has been conducted by determining the availability status of some macro- and micro-nutrients during amelioration of a calcareous saline-sodic soil ($pH_s$ = 8.2–8.6, $EC_e$ = 7.4–9.0 dS $m^{-1}$, SAR = 55.6–73.0). The bioremediation treatments included cropping of sesbania, sordan, or Kallar grass for 15 months. There was an increase in phosphorus (P), zinc (Zn), and copper (Cu) availability in the bioremediation plots resulting from the production of root exudates and likely dissolution of some nutrient-coated calcite. Conversely, the non-cropped gypsum treatment decreased the availability status of these nutrients. Besides leaching losses, adsorption of nutrients on some newly formed $CaCO_3$, a secondary consequence of gypsum dissolution, contributed to this decrease. Soil N content was decreased in all the treatments except for $N_2$-fixing sesbania treatment where N content was increased from 0.49 g $kg^{-1}$ to 0.53 g $kg^{-1}$. There was no treatment effect on soil potassium (K) availability since illite, a K-bearing mineral, was dominant in the clay fraction [15].

Soil microbial biomass is an agent of transformation for added and native organic matter and acts as a labile reservoir for several plant-available nutrients. The activity of microbial biomass is commonly used to characterize microbiological status of a soil and to determine the effects of agricultural practices on soil microorganisms. Dehydrogenase activity (DHA) in soils is related to microbial populations, respiration activity and soil organic matter, and provides an index of the overall microbial activity [16]. This parameter has been studied in few experiments dealing with sodic soil amelioration through chemical and biological means. After using several combinations of chemical and vegetative bioremediation treatments, DHA and microbial biomass carbon (MBC) were determined [17]. The treatments consisted of Karnal grass grown for 1 or 2 years (harvested biomass removed or left to decompose on the soil surface), gypsum application (at 14 Mg $ha^{-1}$) + Karnal grass, gypsum + sorghum, gypsum + rice, and gypsum + sesbania. The soil on which these treatments were applied was alkali ($pH_{1:2}$ = 10.6, $EC_{1:2}$ = 2.1 dS $m^{-1}$, ESP = 95, DHA

= 4.5 μg triphenylformazan $g^{-1}$, MBC = 56.7 mg $kg^{-1}$). The levels of DHA in post-amelioration soil were greater (118.7 μg triphenylformazan $g^{-1}$) in the bioremediation treatments than gypsum + crop treatments (96.1 μg triphenylformazan $g^{-1}$). The MBC values were greater in gypsum + crop treatments (206.3 mg $kg^{-1}$ soil) than in the cropped treatments (161.7 mg $kg^{-1}$ soil). The overall experimental average of MBC (184 mg $kg^{-1}$ soil) for all the treatments was more than three times the initial level of 56.7 mg $kg^{-1}$ soil. In an earlier study [18], a significant increase in urease and dehydrogenase activities was found in alkali soils under permanent vegetation such as grasses. Green manuring of an alkali soil with sesbania has also been reported to increase urease and dehydrogenase activities [19].

Sodic and saline-sodic soils have lost a large fraction of their original carbon (C) pool [20]. The magnitude of the loss may range between 10–30 Mg C $ha^{-1}$, depending on the antecedent pool and the severity of degradation. The soil C pool is not only important for the soil to perform its productivity and environmental functions, but also plays an important role in the global C cycle. In addition to amelioration effect, cultivation of appropriate crops, shrubs, and trees on sodic and saline-sodic soils has the potential to mitigate accelerated greenhouse effects by increasing soil C through biomass production (Tab. 1). Monitoring changes in an alkali soil cropped with four tree species – acacia (*Acacia nilotica* Willd ex Delile), shisham (*Dalbergia sissoo* Roxb. ex DC.), mesquite (*Prosopis juliflora* (Swartz) DC.) and arjuna (*Terminalia arjuna* Bedd.) – suggested shisham and mesquite as more efficient in terms of biomass production and decreasing $Na^{+}$ levels in the soil. Similarly, there was greater microbial activity in upper 0.6 m soil under these species due to the accumulation of humus from decomposition of leaf litter and root decay, which increased soil organic C. The rate of increase was low for the first 2–4 years, exponential between 4–6 years, and plateau at a low rate for 6–8 years [21]. Establishment of mesquite on a sodic field increased organic C of the top 1.2 m soil from 11.8 Mg C $ha^{-1}$ to 13.3 Mg C $ha^{-1}$ in 5 years, 34.2 Mg C $ha^{-1}$ in 7 years, and 54.3 Mg C $ha^{-1}$ in 30 years. The average annual rate of increase in soil organic C was 1.4 Mg $ha^{-1}$ over the 30-year period [22]. Other estimates from field studies on alkali soils suggest that different land-use systems consisting of a number of grasses and trees can sequester organic C in the range of 0.2–0.8 Mg C $ha^{-1}$ $yr^{-1}$ [6].

## Plant species for vegetative bioremediation

The selection of plant species for vegetative bioremediation is generally based on the ability of the species to withstand ambient levels of soil salinity and sodicity while also providing a saleable product or one that can be used on-farm. Considerable variation exists among crops to withstand saline-sodic conditions [23]. Such inter- and intra-crop diversity suggests that field trials be conducted to identify local crops that are adaptable to saline-sodic soil conditions [24]. The farmers, farm advisors, and researchers familiar with local conditions, including crop response to adverse soil conditions and cropping strategies that fit into the local economic conditions, could provide a valuable resource base for making appropriate recommendations. In addition, application of plant breeding approaches is needed to develop crop genotypes with enhanced salt resistance and performance in field conditions [25].

**Table 1.** Potential of two land-use systems (grass only and tree-grass) for carbon (C) sequestration in a calcareous alkali soil (pH = 10.0–10.2; EC = 2.0–6.4 dS $m^{-1}$). Recalculated from [6]

| Treatment[a] | Organic C in soil (g $kg^{-1}$) at different depths[b] | | | C sequestration (Mg $ha^{-1}$ $yr^{-1}$)[c] |
|---|---|---|---|---|
| | 0–0.075 m | 0.075–0.15 m | Mean | |
| *Desmostachya* | 2.9 | 1.6 | 2.3 | 0.33 |
| *Sporobolus* | 2.4 | 1.3 | 1.8 | 0.17 |
| *Acacia* + *Desmostachya* | 3.6 | 1.8 | 2.7 | 0.47 |
| *Dalbergia* + *Desmostachya* | 4.6 | 2.4 | 3.5 | 0.73 |
| *Prosopis* + *Desmostachya* | 4.7 | 2.5 | 3.6 | 0.77 |
| *Acacia* + *Desmostachya* | 2.6 | 1.4 | 2.0 | 0.23 |
| *Dalbergia* + *Desmostachya* | 3.2 | 1.7 | 2.5 | 0.40 |
| *Prosopis* + *Desmostachya* | 3.6 | 1.9 | 2.8 | 0.50 |

[a]*Desmostachya* (*Desmostachya bipinnata* (L.) Stapf), *Sporobolus* (*Sporobolus marginatus* Hochst. ex A. Rich), *Acacia* (*Acacia nilotica* (L.) Delile), *Dalbergia* (*Dalbergia sissoo* Roxb. ex DC), *Prosopis* (*Prosopis juliflora* (Sw.) DC)
[b]After 6 years of plantation
[c]Assuming initial C content in the soil as 1.3 g $kg^{-1}$ (average of the C content, which ranged from 1.0–1.6 g $kg^{-1}$) and mass of 0.15 m depth of 1 ha as $2\times 10^6$ kg, the rate of organic C sequestration in the soil under each treatment was calculated as:
Organic C sequestr. (Mg $ha^{-1}$ $yr^{-1}$) = [(mean C content – original C content in soil) 2] / 6

Several crops, shrubs, trees, and grasses have been used as vegetative bioremediation tools to ameliorate sodic and saline-sodic soils. Some researchers have favored the inclusion of Kallar grass [9], sesbania [11], alfalfa [26], Bermuda grass [8], or sordan [4] as the first crop to accelerate sodic soil amelioration. Several other plant species have produced adequate biomass on salt-affected soils. These include shrub species from the genera *Atriplex* and *Maireana* [27, 28], *Kochia scoparia* L. [29], *Salicornia bigelovii* Torr. [30], *Echinochloa crusgalli* (L.) P. Beauv. [31], and *Portulaca oleracea* L. [32], among others. However, it is imperative to compare them with other species already tested for sodic soil amelioration. In addition, efforts are needed to search other crops such as high-value medicinal and aromatic species with the potential for adequate growth on sodic and saline-sodic soils.

A number of tree plantations have been grown on sodic and saline-sodic soils. These include: *Terminalia arjuna* (Roxb. ex DC.) Wight and Arn. [33], *Prosopis juliflora* (Sw.) DC. [22], *Dalbergia sissoo* Roxb. ex DC., *Acacia nilotica* (L.) Willd. ex Delile [6], *Parkinsonia aculeata* L. and *Prosopis cineraria* (L.) Druce [34], *Sesbania sesban* (L.) Merr. and *Tamarix dioica* Roxb. ex Roth [35], and *Leucaena leucocephala* (Lam.) de Wit [36], among others. In Australia, revegetation by trees was found to be the best long-term option for controlling dryland salinity [37]. Useful information is available regarding sources of seeds, nursery raising techniques, and land preparation and planting procedures for 18 different tree and shrub species with potential for growth on salt-affected soils [34].

Based on cost and benefit analysis, several studies have compared economics of sodic soil amelioration. A net economic loss (cost:benefit 1.00:0.75) was found during vegetative bioremediation although the growth of Karnal grass was adequate, which helped reduce soil sodicity. The economic loss was attributed to the small market demand of the grass in the presence of other good-quality forages in that locality [38]. On the other hand, the bioremediation strategy has been found to be economically beneficial when there was a market demand or local utilization of the crops at the farm level [39, 40]. Agroforestry systems comprising several tree species on saline-sodic soils have been found to be economically feasible in some developing countries because of firewood need in local markets [36]. On the other hand, the market for firewood is not supportive to make agroforestry economically viable in California [8]. Preliminary assessments in Australia suggest that there are 26 salt-resistant plant species capable of producing 13 products (or services) of value to agriculture [27]. From an economic perspective much depends on local needs. In an immediate sense, vegetative bioremediation can only be economically feasible if the selected crops, grasses, or trees have a market demand or local utilization at the farm level. In the long run, one must also consider the value of the improved soils.

## Conclusions

In recent decades, vegetative bioremediation has been found to be an efficient, inexpensive, and environmentally acceptable strategy to ameliorate sodic and saline-sodic soils. Its comparable performance with that of chemical amelioration highlights the effective role of cropping in the amelioration of these soils. Vegetative bioremediation has shown to be advantageous in several aspects: 1) no financial outlay to purchase chemical amendments, 2) accrued financial or other benefits from crops grown during amelioration, 3) promotion of soil-aggregate stability and creation of macro-pores that improve soil hydraulic properties and root proliferation, 4) greater plant-nutrient availability in soil after vegetative bioremediation, 5) more uniform and greater zone of amelioration in terms of soil depth, 6) sequestration of C in post-amelioration soil, and 7) environmentally feasible and productive use of otherwise marginal and degraded soils. However, vegetative bioremediation is slower in effecting positive change than chemical approaches and is contingent on the presence of calcite in soil, which is common when compared to most sodic and saline-sodic soils of arid regions. In addition, its scope becomes limited on highly sodic soils where growth of the bioremediation crops is likely to be variable and patchy and the use of chemical amendments such as gypsum is inevitable. Clearly, vegetative bioremediation is an effective low-cost intervention for resource-poor farmers. This approach has the potential for large-scale adoption under government or community-based programs aimed at the amelioration and improved productivity of degraded common property resources.

## References

1 Ghassemi F, Jakeman AJ, Nix HA (1995) *Salinisation of Land and Water Resources: Human Causes, Extent, Management and Case Studies.* CABI Publishing, Wallingford, UK
2 Gupta RK, Abrol IP (1990) Salt-affected soils: their reclamation and management for crop production. *Adv Soil Sci* 11: 223–288
3 Qadir M, Oster JD (2002) Vegetative bioremediation of calcareous sodic soils: history, mechanisms, and evaluation. *Irrigation Sci* 21: 91–101
4 Robbins CW (1986) Sodic calcareous soil reclamation as affected by different amendments and crops. *Agron J* 78: 916–920
5 Qadir M, Qureshi RH, Ahmad N, Ilyas M (1996) Salt-tolerant forage cultivation on a saline-sodic field for biomass production and soil reclamation. *Land Degrad Develop* 7: 11–18
6 Kaur B, Gupta SR, Singh G (2002) Bioamelioration of a sodic soil by silvopastoral systems in northwestern India. *Agroforestry Systems* 54: 13–20
7 McGrath SP, Zhao FJ, Lombi E (2002) Phytoremediation of metals, metalloids, and radionuclides. *Adv Agron* 75: 1–56
8 Oster JD, Shainberg I, Abrol IP (1999) Reclamation of salt affected soils. In: RW Skaggs, J van Schilfgaarde (eds) In: *Agricultural Drainage.* ASA-CSSA-SSSA, Madison, USA, 659–691
9 Kumar A, Abrol IP (1984) Studies on the reclaiming effect of Karnal-grass and para-grass grown in a highly sodic soil. *Indian J Agric Sci* 54: 189–193
10 Hamid A, Chaudhry MR, Ahmad B (1990) Biotic reclamation of a saline-sodic soil. In: Proceedings of Symposium on Irrigation Systems Management/Research, 73–86. 12–14 November 1990. USAID and Government of Pakistan, Islamabad, Pakistan
11 Ahmad N, Qureshi RH, Qadir M (1990) Amelioration of a calcareous saline-sodic soil by gypsum and forage plants. *Land Degrad Rehabil* 2: 277–284
12 Muhammed S, Ghafoor A, Hussain T, Rauf A (1990) Comparison of biological, physical and chemical methods of reclaiming salt-affected soils with brackish groundwater. In: Proceedings of the 2nd National Congress of Soil Science, 35–42, 20–22 December 1988, Soil Science Society of Pakistan, Faisalabad, Pakistan
13 Qadir M, Ahmad N, Qureshi RH, Qasim SM, Javaid M (1992) Biochemical reclamation of a calcareous saline-sodic soil. *Pakistan J Agric Sci* 29: 406–411
14 Qadir M, Oster JD (2004) Crop and irrigation management strategies for saline-sodic soils and waters aimed at environmentally sustainable agriculture. *Sci Total Environ* 323: 1–19
15 Qadir M, Qureshi RH, Ahmad N (1997) Nutrient availability in a calcareous saline-sodic soil during vegetative bioremediation. *Arid Soil Res Rehabil* 11: 343–352
16 Włodarczyk T, Stępniewski W, Brzezińska M (2002) Dehydrogenase activity, redox potential, and emissions of carbon dioxide and nitrous oxide from Cambisols under flooding conditions. *Biology Fertility Soils* 36: 200–206
17 Batra L, Kumar A, Manna MC, Chhabra R (1997) Microbiological and chemical amelioration of alkaline soil by growing Karnal grass and gypsum application. *Exp Agric* 33: 389–397
18 Rao DLN, Ghai SK (1985) Urease and dehydrogenase activity of alkali and reclaimed soils. *Australian J Soil Res* 2: 661–665
19 Rao DLN, Pathak H (1996) Ameliorative influence of organic matter on biological activity of salt-affected soils. *Arid Soil Res Rehabil* 10: 311–319
20 Lal R (2001) Potential of desertification control to sequester carbon and mitigate the greenhouse effect. *Climatic Change* 51: 35–72

21 Garg VK (1998) Interaction of tree crops with a sodic soil environment: potential for rehabilitation of degraded environments. *Land Degrad Develop* 9: 81–93
22 Bhojvaid PP, Timmer V (1998) Soil dynamics in an age sequence of *Prosopis juliflora* planted for sodic soil restoration in India. *Forest Ecol Manage* 106: 181–193
23 Maas EV, Grattan SR (1999) Crop yields as affected by salinity. In: RW Skaggs, J van Schilfgaarde (eds) In: *Agricultural Drainage*. ASA-CSSA-SSSA, Madison, USA, 55–108
24 Shannon MC (1997) Adaptation of plants to salinity. *Adv Agron* 60: 76–120
25 Pitman MG, Laüchli A (2002) Global impact of salinity and agricultural ecosystems. In: A Laüchli, U Lüttge (eds) *Salinity: Environment–Plants–Molecules*. Kluwer, Dordrecht, The Netherlands, 3–20
26 Ilyas M, Miller RW, Qureshi RH (1993) Hydraulic conductivity of saline-sodic soil after gypsum application and cropping. *Soil Sci Soc Am J* 57: 1580–1585
27 Barrett-Lennard EG (2002) Restoration of saline land through revegetation. *Agric Water Manage* 53: 213–226
28 Malcolm CV (1993) The potential of halophytes for rehabilitation of degraded lands. In: N Davidson, R Galloway (eds) In: *Productive Use of Saline Land*. ACIAR Proceedings 42. Perth, Western Australia, 8–11
29 Garduno MA (1993) Kochia: A new alternative for forage under high salinity conditions of Mexico. In: H Lieth, A Al Masoom (eds) In: *Towards the Rational Use of High Salinity Tolerant Plants* (Volume 1). Kluwer Academic Publishers, The Netherlands, 459–464
30 Glenn E, Hicks N, Riley J, Swingle S (1996) Seawater irrigation of halophytes for animal feed. In: R Choukr-Allah et al. (eds) In: *Halophytes and Biosaline Agriculture*. Marcell Dekker, Inc., New York, USA, 221–236
31 Aslam Z, Saleem M, Qureshi RH, Sandhu GR (1987) Salt tolerance of *Echinochloa crus-galli*. *Biologia Plantarum* 29: 66–69
32 Grieve CM, Suarez DL (1997) Purslane (*Portulaca oleracea* L.): A halophytic crop for drainage water reuse systems. *Plant Soil* 192: 277–283
33 Jain RK, Singh B (1998) Biomass production and soil amelioration in a high density *Terminalia Arjuna* plantation on sodic soils. *Biomass Bioenergy* 15: 187–192
34 Qureshi RH, Barrett-Lennard EG (1998) *Saline Agriculture for Irrigated Land in Pakistan: A Handbook*. Australian Centre for International Agricultural Research, Canberra, Australia
35 Singh B (1989) Rehabilitation of alkaline wasteland on the Gangetic alluvial plains of Uttar Pradesh, India, through afforestation. *Land Degrad Rehabil* 1: 305–310
36 Qureshi RH, Nawaz S, Mahmood T (1993) Performance of selected tree species under saline-sodic field conditions in Pakistan. In: H Lieth, A Al Masoom (eds) In: *Towards the Rational Use of High Salinity Tolerant Plants* (Volume 1). Kluwer Academic Publishers, The Netherlands, 259–269
37 Farrington P, Salama RB (1996) Controlling dryland salinity by planting trees in the best hydrogeological setting. *Land Degrad Develop* 7: 183–204
38 Singh MV, Singh KN (1989) Reclamation techniques for improvement of sodic soils and crop yields. *Indian J Agric Sci* 59: 495–500
39 Chaudhry MR, Abaidullah M (1988) Economics and effectiveness of biological and chemical methods in soil reclamation. *Pakistan J Agric Res* 9: 106–114
40 Sandhu GR, Qureshi RH (1986) Salt-affected soils of Pakistan and their utilization. *Reclam Reveg Res* 5: 105–113

Biosaline Agriculture and Salinity Tolerance in Plants
Edited by M. Öztürk, Y. Waisel, M.A. Khan and G. Görk


# Bio-reclamation of secondary salinized soils using halophytes

Vijendra P.S. Shekhawat[1], Ashwani Kumar[1] and Karl-Hermann Neumann[2]

[1] *Biotechnology Laboratory, Department of Botany, University of Rajasthan, Jaipur-Rajasthan 302004, India*
[2] *Institut für Pflanzenernährung, Abt. Gewebekultur, IFZ, Giessen, Germany*

Soil salinity has become one of the major determinants of global crop productivity. Consequently reclamation of such soils is a most urgent requirement for world food production and for sustainable development. Out of an estimated area of 173 million hectares of totally degraded land in India, approximately 7 million hectares are affected by salinity. Besides naturally occurring saline soils, the secondary salinity that developed due to saline water irrigation is posing a great threat to the perspectives of increasing food and fodder production. In arid and semi-arid regions like Rajasthan (India) decreasing water table and increased use of deep wells, following electrification of villages, has resulted in an increased salinity of irrigation water and consequently in increased salinization. Although, in context with the increasing population, the importance of irrigated agriculture cannot be ignored and excessive saline water irrigation may convert productive soil into unproductive and salinized soil. In western Rajasthan, the area that is affected by the use of such problematic water is some 880 $km^2$. The irrigation water being used there is moderate to highly saline/sodic [1, 2]. As such a large percentage of land is going out for production, year by year, due to saline water irrigation it is a major contributory factor to soil degradation in India [3]. Several halophytic plant species have been tried in the past for their possible use in reclamation of salt-affected soils [4–8]. Besides their positive impact on salt-affected soils, the potential use of some halophytes as forage and as oil seed crops has also been described [9]. However, use of halophytes for soil reclamation is still in an exploratory stage and only a few field studies for bio-reclamation of saline soil using halophytes have been carried out so far [5, 10]. Therefore keeping this fact in mind, the present investigation was undertaken with the objective to study the utilization of halophytes to remove excess salinity added by irrigation.

## Materials and methods

### *Plant species*

*Salsola baryosma* (Roem and Schult), *Haloxylon recurvum* (Moq.) Bunge ex. Boiss, *Suaeda nudiflora* (Willd.) Moq. (all Chenopodiaceae) were studied in the present investigation. Plants were identified according to Bhandari [11] and dried specimens were deposited in the herbarium of the Department of Botany, University of Rajasthan, Jaipur.

### *Experimental site*

Experiments were conducted at Pachkodia village (district of Jaipur), located at the latitude 26° 5'N, longitude 75° 28'E and altitude of 427 m. This area represents the soil and agro-climatic conditions of about two thirds of Rajasthan. The climate of this area is semi-arid with an average rainfall of approximately 500 mm, with more than 80 % of its precipitation during the months of July and August. Temperatures fluctuate widely during the year, ranging from as high as 45°C in summer to 2–4°C in winter.

### *Experimental plots*

14 field plots of 13.5 $m^2$ each were prepared for each treatment. Sowing was carried out by mixing the seeds in the upper 3 cm layer of the soil. A 5 m space was kept between treatments to ensure that there is no seeping of any mineral or water from one treatment to another.

### *Treatments*

$T_1$-field plots planted with *Salsola baryosma.*
$T_2$-field plots planted with *Haloxylon recurvum.*
$T_3$-field plots planted with *Suaeda nudiflora.*

### *Irrigation of the plants*

Irrigation was carried out two times over a period of 3 months. The irrigation comprised of approximately 10–15 cm depth of water in all the plots. The flood irrigation method was used for irrigation, which is of usual practice in the area. Irrigation water used was of the $C_4$-$S_4$ category [12], having high sodium absorption ratio (SAR), high pH value and high electric conductivity (Tab. 1).

### *Soil samples*

Initial soil samples were collected prior to plantation and referred to as 'initial' soil samples. Seed sowing was carried out. Plants were allowed to grow for 3 months and

**Table 1.** Composition of irrigation water used for irrigation at experimental site

| Determination | Value |
|---|---|
| pH | 8.04 |
| EC | 8750.00 $\mu$S.cm$^{-1}$ |
| $K^+$ | 4.80 mg.L$^{-1}$ |
| $Na^+$ | 1845.00 mg.L$^{-1}$ |
| $Mg^{2+}$ | 29.70 mg.L$^{-1}$ |
| $Ca^{2+}$ | 83.20 mg.L$^{-1}$ |
| SAR* | 39.30 |
| $F^-$ | 65.41 mg.L$^{-1}$ |
| $Cl^-$ | 1010.70 mg.L$^{-1}$ |
| $NO_3{}^{2-}$ | 5.48 mg.L$^{-1}$ |
| $SO_4{}^{2-}$ | 1222.30 mg.L$^{-1}$ |

*SAR was calculated following [12]

then soil samples were collected and referred to as 'final' soil samples. Soil samples were taken from every plot and were collected from five different depths, i.e., 0–10, 10–20, 20–30, 30–40 and 40–50 cm.

*Climatic condition during the experimental period*

During the study period the maximum temperature was in the range of 25–41.7° C. Maximum humidity during the experimental period ranged from 52% to almost 100% on some rainy days. Rainfall during the experimental period ranged between 0.4 mm to 60.4 mm.

*Plant analysis*

The total aerial shoot was cut at the soil line and then dried at 105° C until the weight became constant. Fresh weight and dry weight were determined. Subsequently, dried aerial shoot was ground in a coffee mill up to 1 mm size. 100 mg plant material was placed in porcelain crucibles and ashing was carried out for approximately 20 h at 550° C in a muffle furnace until the organic matter completely disappeared. Ashes were then digested in 50 % v/v nitric acid and diluted with distilled water for analysis. Potassium and sodium were determined using flame emission and $Mg^{2+}$ and $Ca^{2+}$ were determined using atomic absorption.

*Soil and water analysis*

The pH and EC were determined in 1:2.5 (soil:water) extract where extracts were prepared without using vacuum or pressure. For determination of cations, extracts were prepared following Mehlich [13], while using barium chloride as exchanger. Exchangeable sodium, exchangeable potassium, exchangeable calcium and exchangeable magnesium were determined using atomic absorption spectrophotometry (Perkin Elmer model 2380). Exchangeable sodium percentage was calculated following Richards [12].

## Results

*Plant analysis*

Table 2 presents the ion composition and biomass production in plants undertaken in the present investigation.

**Table 2.** Ion accumulation, $Na^+$ uptake and biomass production in *S. baryosma*, *S. nudiflora* and *H. recurvum* over a period of 3 months

| Plant species | Ion accumulated mg.g$^{-1}$.dry wt. | | | | Sodium uptake g.plant$^{-1}$ | Shoot biomass produced kg.dry wt.ha$^{-1}$ |
|---|---|---|---|---|---|---|
| | $K^+$ | $Na^+$ | $Mg^{2+}$ | $Ca^{2+}$ | | |
| *S. baryosma* | 3.44 | 68.16 | 5.59 | 9.21 | 9.61 | 1,847 |
| *S. nudiflora* | 5.94 | 89.86 | 5.09 | 6.16 | 15.63 | 2,175 |
| *H. recurvum* | 5.94 | 67.59 | 5.60 | 5.34 | 17.03 | 2,192 |

*Effect of halophyte plantation on soil characteristics*

*Soil reaction (pH)*
The pH of the soil was considerably modulated by halophyte plantation (Tab. 3). In *S. baryosma* plots a considerable decrease in soil pH was observed in all the depth levels. However in *S. nudiflora* plots, an increase in soil pH was recorded in 10–20 and 20–30 cm soil layers. However soil pH was decreased in 0–10 and 30–50 cm soil layers. In *H. recurvum* plots, soil pH decreased considerably, in all the depth levels. In control plots an increase in pH was recorded in all depth levels.

*Soil salt content (EC)*
Soil salt content markedly differed in initial and final soil samples as indicated by soil electric conductivity (Tab. 3). In *S. baryosma* planted plots a considerable decrease in soil EC was observed in 10–40 cm soil layers. However an increase in soil EC was recorded in 0–10 and 40–50 cm soil layers. In *S. baryosma* plots maximum reduction in soil EC was recorded in 10–20 cm soil layer, which was followed by 20–30 cm and 30–40 cm soil layers. In *H. recurvum* grown field plots, soil electric conductivity (EC) decreased in all depth levels. The reduction in soil EC was 56% to 85% in different depth levels. Maximum reduction (85%) in EC was recorded in 10–20 cm and minimum reduction (56%) was in the 40–50 cm depth layer. In *S.*

**Table 3.** Effect of halophyte plantation on some physical and chemical characters of secondary salinized soil. Final values with different letter (a, b, c) differs significantly from initial values ($P < 0.05$) according to T-test

| | | *S. baryosma* plots | | *S. nudiflora* plots | | *H. recurvum* plots | | Control plots | |
|---|---|---|---|---|---|---|---|---|---|
| Determination | Depth | Initial | Final | Initial | Final | Initial | Final | Initial | Final |
| | 0–10 | 8.16 a | 8.13 a | 8.13 a | 8.04 b | 8.44 a | 7.73c | 8.10 a | 8.42 b |
| pH | 10–20 | 8.29 a | 8.03 b | 7.99 a | 8.04 a | 8.27 a | 7.96c | 8.08 a | 8.33 b |
| | 20–30 | 8.48 a | 7.88 c | 7.97 a | 8.02 a | 8.27 a | 8.01c | 7.90 a | 8.16 b |
| | 30–40 | 8.32 a | 7.90 c | 7.99 a | 7.97 a | 8.38 a | 7.95c | 7.95 a | 7.96 a |
| | 40–50 | 8.27 a | 7.65 c | 7.93 a | 7.82 b | 8.01 a | 7.74c | 7.88 a | 7.89 a |
| | 0–10 | 768 a | 967 b | 930 a | 199 c | 1155 a | 314 c | 868a | 852.5 b |
| EC $_{(1:2.5)}$ | 10–20 | 1068 a | 697 c | 1142 a | 167 c | 1880 a | 273 c | 665 a | 710. b |
| ($\mu S.cm^{-1}$) | 20–30 | 1002 a | 707 c | 1092 a | 167 c | 1006 a | 228 c | 448 a | 639.5 b |
| | 30–40 | 971 a | 754 c | 949 a | 175 c | 939 a | 246 c | 421 a | 629.5 b |
| | 40–50 | 858 a | 1200 b | 585 a | 233 c | 648 a | 279 c | 421 a | 581.0 b |
| | 0–10 | 33.75 a | 97.25 c | 84.75 a | 47.75 b | 74.25 a | 33.75 c | 64.50 a | 104.25 b |
| $Na^+$ | 10–20 | 64.50 a | 90.75 b | 45.00 a | 63.75 b | 80.00 a | 37.75 b | 74.50 a | 98.25 b |
| ($mg.100g^{-1}$) | 20–30 | 69.75 a | 89.25 b | 75.25 a | 76.25 a | 81.25 a | 35.00 b | 79.00 a | 98.25 b |
| | 30–40 | 73.50 a | 96.25 b | 78.50 a | 81.25 a | 77.23 a | 39.00 c | 80.00 a | 92.00 b |
| | 40–50 | 66.00 a | 106.50 b | 70.25 a | 92.25 b | 64.00 a | 42.75 b | 63.50 a | 99.25 b |
| | 0–10 | 12.00 a | 17.00 c | 7.75 a | 22.25 b | 5.50 a | 15.75 c | 19.50 a | 11.50 b |
| $Ca^{2+}$ | 10–20 | 13.25 a | 6.50 b | 11.00 a | 19.50 b | 8.50 a | 19.00 b | 16.50 a | 10.50 b |
| ($mg.100g^{-1}$) | 20–30 | 14.25 a | 8.00 b | 9.75 a | 18.00 b | 10.25 a | 14.50 b | 11.75 a | 9.75b |
| | 30–40 | 15.75 a | 7.00 b | 13.00 a | 19.00 b | 12.00 a | 10.50 c | 12.75 a | 8.75b |
| | 40–50 | 14.00 a | 12.00 a | 11.00 a | 18.00 b | 11.75 a | 13.50 b | 13.25 a | 10.00b |
| | 0–10 | 50.58 | 77.95 | 81.8 | 55.31 | 80.79 | 46.01 | 62.3 | 79.0 |
| ESP | 10–20 | 70.35 | 87.18 | 75.8 | 64.57 | 79.47 | 44.63 | 60.5 | 80.6 |
| | 20–30 | 73.09 | 84.59 | 74.5 | 70.84 | 77.99 | 46.44 | 66.1 | 80.09 |
| | 30–40 | 73.86 | 86.30 | 75.6 | 70.35 | 74.79 | 51.73 | 76.6 | 80.20 |
| | 40–50 | 72.70 | 82.64 | 63.5 | 74.13 | 71.56 | 50.61 | 66.1 | 76.07 |

*nudiflora* plots a significant decrease in soil EC was also recorded at all depth levels. Reduction in soil EC ranged from 60% to 85% in different depth levels. Maximum reduction (85%) in soil EC was recorded for 10–20 cm soil layer, which was followed by 20–30 cm soil layer where 84% reduction for soil EC was recorded. In *S. nudiflora* grown field plots, least reduction in soil EC was recorded in 40–50 cm soil layer. In control plots, where no halophytes were grown, an increase in soil EC was observed in 20–50 cm soil layers. However a decrease in soil EC was recorded in the upper 0–10 cm soil layer.

*Soil exchangeable sodium percentage (ESP)*

Plants of *S. baryosma* were not able to reduce ESP and an increase in ESP was recorded in all depth levels. Maximum net increase in ESP was recorded in the upper 0–10 cm soil layer. In *S. nudiflora* planted plots, soil ESP decreased considerably in 0–40 cm depth soil layers. However plants failed to reduce soil ESP in 40–50 cm soil layer and increase in soil ESP was recorded for this depth level. In *H. recurvum* plots, a remarkable decrease in soil ESP was recorded in all depth levels.

*Soil exchangeable sodium and calcium*

An increase in soil exchangeable $Na^+$ was recorded in *S. baryosma* plots (Tab. 3). Maximum increase in soil exchangeable $Na^+$ was recorded in the upper 0–10 cm soil layer. In *S. nudiflora* grown plots, the amount of exchangeable $Na^+$ was markedly reduced in the upper 0–10 cm soil layer. In other soil layers, i.e., 10–50 cm depth, an increase in the amount of exchangeable $Na^+$ was recorded. However, in 20–30 and 30–40 cm soil layers, only marginal increases in soil exchangeable $Na^+$ were recorded. In *H. recurvum* grown plots, the amount of exchangeable $Na^+$ was reduced considerably in all depth levels. Maximum reduction in exchangeable $Na^+$ was recorded in 20–30 cm soil layer. In control plots, an increase in exchangeable $Na^+$ was noticed in all depth levels. In *S. nudiflora* plots a significant increase in soil $Ca^{2+}$ content was recorded at all depth levels. Soil $Ca^{2+}$ content in *H. recurvum* plots increased significantly in the 0–30 cm soil layer. In the 40–50 cm depth level a marginal increase in soil exchangeable $Ca^{2+}$ was also recorded. However in the 30–40 cm depth level a decrease in soil $Ca^{2+}$ content was noticed. In *S. baryosma* grown plots, soil exchangeable $Ca^{2+}$ content increased only in the 0–10 cm soil layers. However a significant decrease in soil exchangeable $Ca^{2+}$ content was recorded in 10–50 cm soil layers.

## Discussion

Our findings demonstrated that all three species of halophytes can be utilized as 'primer plants' and for phyto-remediation of secondary salinized agricultural fields. All three species had a clear modulatory effect on different soil physical and chemical properties.

Plants may influence the soil physical properties like pH and EC [14] and by that may counter the effect of salinity/alkalinity. *H. recurvum* and *S. nudiflora* were superior when compared to *S. baryosma* in reducing EC. The electric conductivity of the soil extracts of all the plots was far above the electric conductivity values of those found in Central Europe [15]. However, after plantation with halophytes these values can be reduced. Similar results for reduction in soil relative electric conductivity (REC) of saline-sodic soil (mainly above a 45 cm depth) by *Echinochloa stagninum* was reported by Helalia et al. [5]. Positive results for reduction of total soluble solids (TSS) by plantation of halophytic species *Juncus acutus* and *Juncus rigidus* were reported by Zahran et al. [7]. The improvement in soil permeability due to root action may also facilitate leaching, which in turn causes reduction in the EC of the upper soil layers. The soil pH is directly affected by the concerns of plant roots to $H^+$, $OH^-$, $HCO_3^-$, and organic anions reactions at soil exchangeable complex in response to disequilibria in cation and anion uptake effective under any particular soil system [16]. In addition to root respiration, root exudates, increased microbial activity and organic matter added by vegetation may alter the soil solution quality, which may influence the soil pH. Release of acidic root exudates decreases the soil pH [17]. Furthermore, $NH_4^+$ uptake by the plants may also reduce the soil pH significantly [18, 19]. Reduction in soil pH by growing *Leptochloa fusca* has been reported by Qadir et al. [6]. Extrusion of $H^+$ ion from roots, following ion transport mechanisms, is a general phenomenon in plant roots under a saline environment [20] and may contribute significantly to reduction in pH. In the present investigation, soil pH was considerably modulated by the planted halophytes. The effect of these plants on soil pH was variable at different depth levels. Although dependence of soil pH upon different depth levels under different conditions like agriculture, forestry and natural conditions have been clearly established [16, 21], it may also be attributed to different root morphologies of the different plants investigated. The more accentuated influence of *H. recurvum* could be related to its deep reaching root system as compared to shallower root systems of *S. nudiflora* and *S. baryosma*. Chaudhary et al. [22] discussed the ability of *Suaeda fructicosa* to accumulate $Na^+$ and other ions. A single plant of *Suaeda fructicosa* may accumulate some 100 g of salt in its aerial tissue. In the present investigation, maximum amount of $Na^+$ was accumulated during a period of 3 months by *H. recurvum* followed by *S. nudiflora* and *S. baryosma*. Reduction in sodium content at the 20–30 cm depth level by plantation of *Suaeda salsa* plants has been reported by Zhao [10]. The root action of halophytes may mobilize the native lime of soil. Robbins [23] reported that $CO_2$ released during root respiration might be a major contributing factor for the reclamation of salt affected lands. Formation of $Ca(HCO_3)_2$ from $CaCO_3$ may also take place in the presence of $H_2CO_3$ resulting in increased solubility of $CaCO_3$ [24]. The $Ca^{+2}$ thus released may replace the $Na^+$ from the exchange sites of the soil colloidal complex. The replaced $Na^+$ together with excess salts may be washed away from the root zone by rainwater or by any other source. In the present investigation *S. nudiflora* was the most efficient plant in increasing soil exchangeable calcium content. Further concept of biopores, i.e., pores left behind in the soil after death or decay of halophyte roots, may also be quite conceivable here [25]. Although such pores are in the macropore category ($> 100\mu m$ diameter) and are created by thick roots, they are large enough to provide channels for optimum water and air conduction [25]. Consequently, roots of halophytes may alter several soil physical and chemical characteristics. As salt affected soils are generally degraded structurally and have a low permeability, growing halophytes may improve such concerns regarding soil–water and soil–air relations. All three halophyte species had a considerable impact on soil quality. The overall efficiency of these halophytes in reclamation was in the decreasing order of *H. recurvum* > *S. nudiflora*> *S. baryosma*. Plantation of these halophytes considerably altered the soil pH, EC, exchangeable $Na^+$, exchangeable $Ca^{2+}$ and exchangeable sodium percentage (ESP). However, variability at different depth levels needs closer examination and continuous study. Investigations of soil structure and a closer characterization of differences in the root system of the three species should be further investigated. Halophytes may accelerate the reclamation process of salinization that otherwise would be unproductive or poor in yield. More than 1 year of intercropping will certainly be required to reclaim such soils and success surely will be influenced by the efficiency of rainfall. However, planting halophytes could replace fallowing for 1 or 2 years, as practiced presently, bringing more benefits for saline soils.

## References

1 Gupta IC (1990) *Use of saline water in agriculture: A study of arid and semi arid zones in India.* Revised edition. Oxford and IBH, New Delhi

2 Singh N, Kolarkar AS, Bohra PC (1994) Quality of ground water and its effect on soil properties in Samdari-Siwana-Balotra areas of Western Rajasthan. *Ann Arid Zone* 33: 287–293

3 Sehgal JL, Sexena RK, Pofali RM (1990) Degraded soils, their mapping through soil surveys. In: IP Abrol, VV Dhruvanarayna (eds) In: *Technologies for wasteland development*. Pub. Inform. Div. ICAR, New Delhi, 1–20

4 DeVilliers AJ, VanRooyen MW, Theron GK, Claassens AS (1995) Removal of sodium and chloride from a saline soil by *Mesembrayanthemum barklyi*. *J Arid Environ* 29: 325–330

5 Helalia AM, El-Amir S, Abou–Zeid ST, Zaghloul KF (1992) Bio-reclamation of saline sodic soil by Amshot grass in Northern Egypt. *Soil Till Res* 22: 109–116

6 Qadir M, Qureshi RH, Ahmad N (1996) Reclamation of a saline-sodic soil by gypsum and *Leptochloa fusca*. *Geoderma* 74: 207–217

7 Zahran MA, Amal A, Wahid A (1982) Halophytes and human welfare. In: DN Sen, KS Rajpurohit (eds) In: *Tasks for vegetation science, Contributions to the Ecology of Halophytes*. Dr. W. Junk Publishers, The Hague

8 Yadav JSP (1975) Improvement of saline alkali soils through biological methods. *Indian For* 385–395

9 Glenn EP, Brown JJ, Blumwald E (1999) Salt tolerance and crop potential of halophytes. *Cri Rev Plant Sci* 18: 227–255

10 Zhao KF (1991) Desalinization of saline soils by *Suaeda salsa*. *Plant Soil* 135: 303–305

11 Bhandari MM (1990) *Flora of the Indian desert*. Revised edition. M P S Repros, Jodhpur

12 Richards LA (ed.) (1954) Diagnosis and improvement of saline and alkaline soils. Agriculture. USDA agriculture handbook 60. US Govt Print Office, Washington DC

13 Mehlich A (1942) Rapid estimation of base-exchange properties of soil. *Soil Sci* 53: 8–14

14 Marschner H (1995) Mineral nutrition of higher plants, $2^{nd}$ Ed. Academic Press, London

15 Neumann KH, Pauler B (2000) Fallstudie zur Bedeutung erhöhter Salzkonzentrationen im Beregnungswasser unter den humiden Bedingungen Mitteleuropas. URL:http://ibid.uni-giessen.de/ghtm/2000/uni/p000005

16 Helyar KR, Porter WM (1989) Soil acidification, its measurement and the processes involved. In: AD Robson (ed.) In: *Soil Acidity and Plant Growth*. Academic Press, Sydney, Australia, 61–101

17 Dormaar JF (1988) Effect of plant roots on chemical and biochemical properties of surrounding discrete soil zones. *Can J Soil Sci* 68: 233–242

18 Nye PH (1981) Changes of pH across the rhizosphere induced by roots. *Plant Soil* 61: 7–26

19 Gorham A, Wyn Jones RG, McDonnel E (1985) Some mechanism of salt tolerance in crop plants. *Plant Soil* 89: 15–40

20 Soon YK, Miller MH (1977) Changes in the rhizosphere due to $NH_4$ and $NO_3$ fertilization and phosphorus uptake by corn seedlings (*Zea mays* L.). *Soil Sci Soc Am J* 41: 77–80

21 Rengasamy P, Chittleborough D, Heylar K (2003) Root zone constraints and plant based solutions for dry land salinity. *Plant Soil* 257: 249–260

22 Chaudhri I, Shah BH, Naqvi NI, Mallik IA (1964) Investigation on the role of *Suaeda fructicosa* Forsk. in the revegetation of saline and alkali soil in West Pakistan. *Plant Soil* 21: 1–7

23 Robbins CW (1986) Carbon dioxide partial pressure in lysimeter soils. *Agron J* 78: 151–158

24 Kovda VA, Van den Berg C, Hagen RM (1973) *Irrigation, Drainage, and Salinity: An international source book*. FAO/UNESCO, Hutchinson and Co, London

25 Yunusa IAM, Newton PJ (2003) Plants for amelioration of subsoil constraints and hydrological control: the primer – plant concept. *Plant Soil* 257: 261–281

Biosaline Agriculture and Salinity Tolerance in Plants
Edited by M. Öztürk, Y. Waisel, M.A. Khan and G. Görk


# Effects of irrigation water salinity on yield and evapotranspiration of drip irrigated cucumber in a semiarid environment

Halil Kirnak

*University of Harran, Agriculture Faculty, Irrigation Department, 63200 Sanliurfa, Turkey*

## Introduction

The major limiting factor on the expansion of irrigated agriculture throughout the world is the lack of water. Water demand is increasing due to fast population growth rates, improvement in living standards, improvement in industry and municipality, and global warming. Nowadays, there is an increasing tendency to use saline irrigation water in arid and semi-arid regions of the world because of rising water demands for irrigation. Slightly and moderately saline water can be used for irrigation successfully to grow salt tolerant and moderately salt-tolerant crops without adverse long-term effects on soil provided appropriate soil water management practices are followed [1].

Crop evapotranspiration (ET) under standard conditions applying different soil moisture regimes were studied but rarely under saline irrigation water. In general, salinity and drought affect the plant in a similar way. With increasing salinity or drought, soil water availability decreases. However, a question has arisen whether the yield-ET model developed under drought conditions can be valid under salinity conditions.

Stewart et al. [2] demonstrated that the relationship between yield and ET of maize was the same in cases of drought and salinity. Katerji et al. [8] checked this hypothesis for sunflower, potatoes and soybean in Italy using saline water. They reported that the hypothesis developed by Steward et al. [2] was valid for sunflower and, to a lesser degree for potatoes, but not for soybean. Shalhevet [3] reported that crop water production functions relating yield to evapotranspiration are not influenced by water salinity. Shalhevet and Hsiao [4] studied the effects of salinity and drought on cotton and pepper. They concluded that, at the same soil water potential, plants grown under saline conditions showed better growth than under drought. It seems that the yield response to ET due to salinity or drought is still a controversial subject.

The yield-ET relationship developed by Stewart et al. [2] was:

$$\left(1 - \frac{Y_a}{Y_m}\right) = k_y \left(1 - \frac{ET_a}{ET_m}\right) \tag{1}$$

Where: $Y_a$ = actual harvested crop yield, $Y_m$ = maximum harvested crop yield, $k_y$= yield response factor, $ET_m$ = maximum ET, $ET_a$ = actual ET.

Cucumber (*Cucumis sativus*) is considered to be a moderately salt-sensitive crop. Most of the studies in the literature have been conducted for determining the effect of saline water application on fruit quality and yield of field crops, vegetables and orchards rather than cucumber. On the other hand, the response of cucumber under saline irrigation water has not been adequately characterized. This research was conducted: 1) to check the hypothesis predicting yield from ET under saline conditions with models developed for drought conditions, 2) to quantify crop growth and yield of cucumber when grown under different irrigation water salinity levels, and 3) to determine threshold EC of irrigation water for drip irrigated cucumber.

## Materials and methods

### *Plant culture, treatments and irrigation*

The experiment was carried out on a clay loam soil, classified as Ikizce soil series (*Vertic Calciorthid aridisol*), from April–July 2001 in Sanliurfa, Turkey. The altitude, latitude and longitude of the experimental site are 465 m, 37°08'N and 38°46'E, respectively. The weather is hot and dry from May to September, when temperatures can reach up to 46°C. The 0–60 cm depth of soil profile had a dry bulk density of 1.32 g/cm$^3$, pH of 7.1, and $EC_e$ of 1.0 dS/m. The field capacity (FC) was 32.50 %, and the permanent wilting point (PWP) 21.60 %, as determined gravimetrically. Irrigation water was of good quality with $EC_i$ of 0.45 dS/m, containing (meq/L) 1.1 $Ca^{2+}$, 1.0 $Mg^{2+}$, 0.25 $Na^+$, 0.02 $K^+$, 0.75 $SO_4{}^{2-}$, 0.90 $HCO_3{}^-$, 0.60 $Cl^-$ and a pH of 7.0.

A hybrid cultivar 'Beith Alpha F1' of cucumber widely cultivated in southeast Turkey was selected. Seeds were germinated in fine sand during the second week of March and at the second true leaf stage (20 days) similar sized seedlings were transplanted into plastic tubs containing previously washed sand. Similar sized seedlings were again selected at the fourth true leaf stage (12 days) and transplanted to the field in the second week of April. The plants were drip irrigated according to their scheduled program after transplanting.

33 plants per replicate were planted in rows with an inter-plant spacing of 0.5 m and an inter-row spacing of 1.0 m. A single drip irrigation tube (Goktepe Co., Izmir, Turkey), with 4.0 L/h and 0.5 m emitter spacing, was placed for each row. Each experimental plot was composed of three 5 m long rows. Yield was measured only for plants growing in the central row, the outer rows serving as borders. The operating pressure of the drip irrigation system was constant during the experiment as 100 kPa. Each plot had a separate flow meter (Teksan Co., Turkey) to monitor water input.

All treatments received the same amounts of total N (12 kg/da), $P_2O_5$ (15 kg/da)and $K_2O$ (24 kg/da) fertilizers. Based on soil test results, all of the P as a di-ammonium phosphate (18-46-0) was applied prior to planting and mixed into the soil. The N as ammonium nitrate (33 % N) and K as potassium nitrate (13-0-46) were added equally at weekly intervals through the drip irrigation system, starting after transplanting until the second harvest. Hand weeding was carried out 3 times during the growing season.

Treatments were: 1) fully irrigated with good quality water of 0.45 dS $m^{-1}$ $EC_i$ (C), 2) fully irrigated with saline water of 3 dS $m^{-1}$ $EC_i$ ($SW_1$), 3) fully irrigated with saline water of 6 dS $m^{-1}$ $EC_i$ ($SW_2$), and 4) fully irrigated with saline water of 9 dS $m^{-1}$ $EC_i$ ($SW_3$). All treatments were drip-irrigated every 3 days. Saline solutions were obtained by adding NaCI to irrigation water to obtain the $EC_i$ levels of 3, 6, 9 dS $m^{-1}$.

Evapotranspiration (ET) for each plot was calculated according to the water balance approach. In order to determine the actual ET, soil moisture content between 0 and 90 cm was measured gravimetrically prior to irrigation.

### *Plant growth*

Plant growth was monitored using leaf area measurements. Randomly chosen three whole plants from each plot were sampled at the end of the first harvest. Total leaf area was determined with a portable leaf area meter (LI-3100, LI-COR, Lincoln, NE). The harvesting was initiated in the first week of July. Fruits were harvested every 2–3 days depending on fruit size until the end of July. Leaf relative water content (LRWC) was calculated based on the methods from Yamasaki and Dillenburg [5].

### *Crop salt tolerance and statistical analysis*

The yield response to salinity was evaluated by a linear regression model (threshold-slope model) proposed by Maas and Hoffman [6]:

$$Y_{\mathrm{r}} = 100 - s(EC_{\mathrm{i}} - EC_{\mathrm{t}}) \tag{2}$$

Where $Y_r$ is the percentage of the yield of the crop grown under saline conditions relative to that obtained under non-saline (control) conditions, the $EC_t$ is the threshold salinity value tolerated by the crop without yield loss, the s is the yield loss per unit increase in salinity (or shortly, slope of the regression between relative yield and $EC_e$).

Mass and Hoffman [6] summarized the relationship between salinity and yield as:

$$Y_r = \left\{ \begin{array}{cc} 100 & 0 \le ECe \le EC_t \\ 100 - s\,(EC_e - EC_t) & EC_t \le ECe \le EC_0 \\ 0 & ECe \ge EC_0 \end{array} \right\} \tag{3}$$

Where $EC_0$ is the level of salinity at which the yield is zero.

The layout of the experiment was in a randomized complete-block design with three replications. The data were subjected to ANAVO and Duncan's least significant difference (LSD) test to check the significance.

## Results and discussion

### *Evapotranspiration and yield*

The yield and ET values of each treatment are presented in Table 1. Figure 1 shows the measured and estimated yield of cucumber. The yield estimation of cucumber is very good since the slope and the intercept of the regression line is not significantly different from 1 and 0, respectively. Equation 1 tends to overestimate yield for $SW_1$ and $SW_2$, but underestimated yield for $SW_3$. However, both overestimation and underestimation never exceeds 10 % within the range of measured yields. The linear regression analysis between measured and estimated yields was Y(estimated) = 1.07 * Y (actual) – 0.086 with an $R^2$ of 0.98. The results of this study confirm those of Stewart et al. [2] on maize, Katerji et al. [1] on sunflower and potatoes and support their conclusion concerning a similar relationship between yield and ET for both drought and salinity.

**Table 1.** Effects of irrigation water salinity on fruit yield, ET and plant growth of drip irrigated cucumber grown in semiarid conditions

| Tr. | App. water (mm) | ET (mm) | Total yield (kg $da^{-1}$) | Fruit length (cm) | Fruit diameter (cm) | Fruit weight (g) | Fruit no per plant | Leaf area ($cm^2$) | LRWC (%) |
|---|---|---|---|---|---|---|---|---|---|
| C | 825 | 841 | 3187 a* | 17.5 a | 4.5 a | 235 a | 15.2 a | 1875 a* | 91.1 a |
| $SW_1$ | 796 | 795 | 3067 a | 16.2 a | 4.4 a | 222 a | 14.8 a | 1827 a | 89.2 a |
| $SW_2$ | 714 | 680 | 2311 b | 9.4 b | 2.9 b | 166 b | 6.1 b | 1101 b | 70.2 b |
| $SW_3$ | 402 | 338 | 915 c | 5.8 c | 2.8 b | 76 c | 1.6 c | 412 c | 52.5 c |

*Within each column, means followed by the same letter indicates no significant difference between treatments by Duncan's multiple range test at $P \leq 0.01$

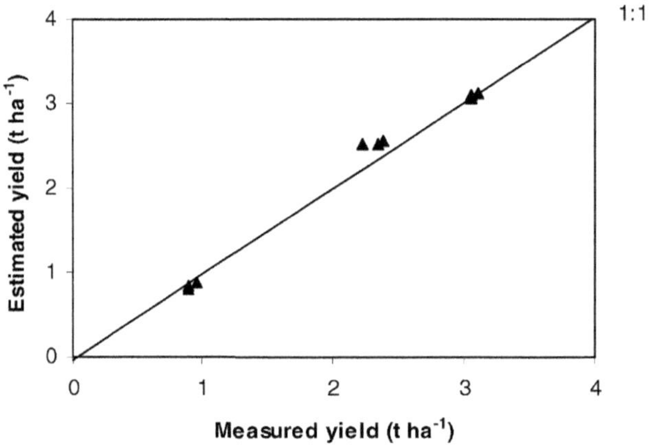


**Figure 1.** Measured yield of cucumber *versus* yield estimated with Equation (1)

Saline irrigation caused an increase of the soil water content due to increased osmotic potential of soil nutrient solution. While the ET of the control treatment was 841 mm, saline treatments $SW_1$, $SW_2$ and $SW_3$ had a 795, 680 and 338 mm of ET, respectively. This is in agreement with the findings of Sonneveld and Voogt [7], who reported that increasing irrigation water salinity reduces transpiration and increases drainage for a given irrigation volume.

Applied water for each treatment was different since irrigation was scheduled to increase soil moisture to field capacity. The applied water and ET for C treatment was 825 and 841 mm, respectively. As the EC of irrigation water increased, the applied irrigation water decreased due to reduced ET. Reduction in ET in the presence of salinity is often partially caused by reduced plant size and fraction of ground cover. The difference observed in ET is a reflection of differences in transpiration. The saline conditions in the root zone hinders water uptake because of reduced osmotic potentials caused by increased salinity, which ultimately decreases transpiration of the crop. The changes among EC of irrigation water, yield and ET are shown in Figure 2. Figure 2 shows that there was a similar trend of reduction for both ET and yield with increasing irrigation water salinity.

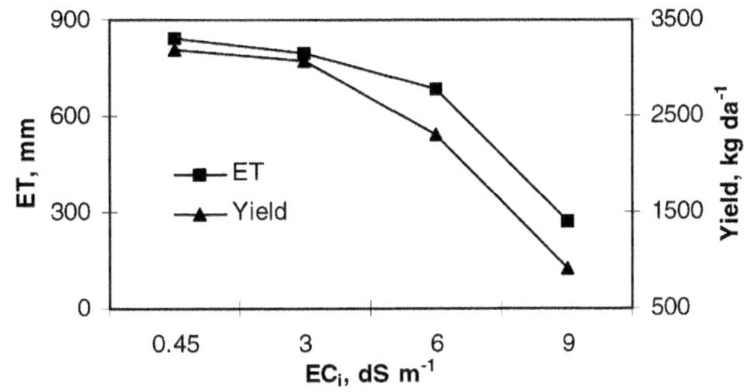


**Figure 2.** Relationship between $EC_i$, yield and ET

Drip irrigation systems helped to maintain higher soil moisture, resulting in higher transpiration. High water content may alleviate inhibition in water uptake caused by salinity. However, this situation is valid to a certain point. Under severe $EC_i$ conditions, crops cannot make necessary internal adjustment of osmotic potential, so ET and plant growth decreased sharply.

The average fruit yield was greater at the control and 3 dS $m^{-1}$ $EC_i$. At 6 and 9 dS $m^{-1}$ $EC_i$, yield was reduced by 27.5 % and 71 %, respectively. Both fruit weight and fruit number were reduced significantly with increasing salinity. An average of 60 % and 89 % fewer fruits were harvested at 6 and 9 dS $m^{-1}$ $EC_i$, respectively, compared to the control treatment. The individual fruit weight at the same irrigation water salinity levels was reduced 29 % and 68 %, respectively, compared to the control treatment.

*Plant growth*

The decline in cucumber vegetative growth at high salinities was expressed as reduced total leaf area. It was found that the maximum leaf area resulted from the lowest water salinity. As the irrigation salinity levels increased, plant growth (leaf area) decreased linearly (Tab. 1). These observations are similar to the results for cotton presented by Vulkan-Levy et al. [8]. Total leaf area reduced an average reduction of 3 %, 41 %, and 78 % with salinity levels of 3, 6 and 9 dS $m^{-1}$, respectively.

The LRWC decreased with increasing salinity of the irrigation water (Tab. 1). The LRWC at waters of 6 and 9 dS $m^{-1}$ $EC_i$ was decreased 23 % and 42 %, respectively, compared to C. This result is in agreement with the findings of Katerji et al. [9] and Maggion et al. [10].

*Crop salt tolerance*

The threshold irrigation salinity value $EC_{i(t)}$ for drip irrigated cucumber was found at 3.4 dS $m^{-1}$ and the $EC_{i(0)}$, which the yield is zero, was 12.3 dS $m^{-1}$. The slope (s) was 11.17 % per dS $m^{-1}$ (Fig. 3). The yield response to irrigation salinity was calculated as Yr = 100 – 11.17 ($EC_i$- 3.4).

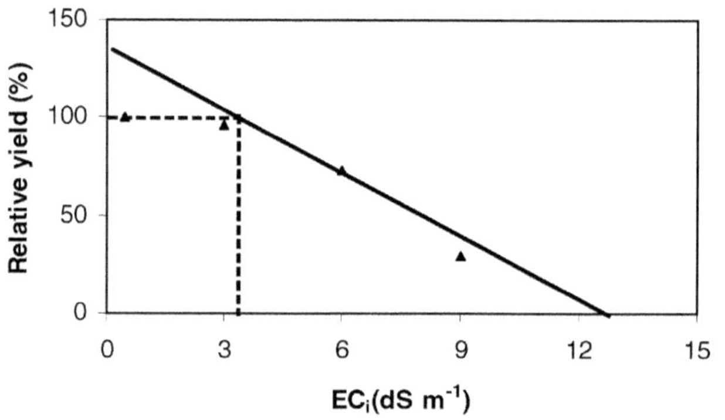


**Figure 3.** Relative yield response of drip irrigated cucumber to increasing salinity of irrigation water

According to Ayers and Westcot [11, $EC_{i(t)}$, $EC_{i(0)}$, and s values for cucumber were 1.7, 6.8 dS $m^{-1}$ and 19.5 % per dS $m^{-1}$, respectively. We calculated that threshold and zero yield values of cucumber were bigger than that of Ayers and Wescot [11], whereas the slope value was lower. These differences could be due to drip irrigation systems which absolutely have some advantages in the use of low quality water. Secondly, we conducted the experiment over only 1 year. The negative effect of the saline irrigation water would occur potentially in the next year. Chartzoulakis [12] found that the threshold irrigation salinity for greenhouse grown cucumber was 1.3 dS $m^{-1}$. The researcher also noted that each unit of EC greater than the threshold decreased yield by 15.9 %.

High temperature and low relative humidity are two main characteristics of arid climate and actually these two parameters increased the ET and likely intensed the negative effect of salinity on crop growth. The arid and semi-arid conditions tend to lower the crop's threshold for salinity stress either because of high transpiration or changes in leaf biochemistry [13]. However, our results showed the opposite of this idea. The likely reason of it could be the irrigation method chosen.

The threshold soil salinity value $EC_{e(t)}$ for drip irrigated cucumber was found to be 1.5 dS $m^{-1}$ and the $EC_{e(0)}$, which the yield is zero, was 8.85 dS $m^{-1}$. The slope (s) was 13.6 % per dS $m^{-1}$ (Fig. 4). The yield response to soil salinity was calculated as $Yr = 100 - 13.6\,(EC_e - 1.5)$. Allen et al. [14] reported $EC_{e(t)}$ between 1.1–2.5 dS $m^{-1}$ and slope between 7–13 % per dS $m^{-1}$ based on climate, soil conditions and cultural practice. Although our $EC_e$ threshold value was between reported values of Allen et al. [14] , the slope value was a little bit higher than that of Allen et al. [14].

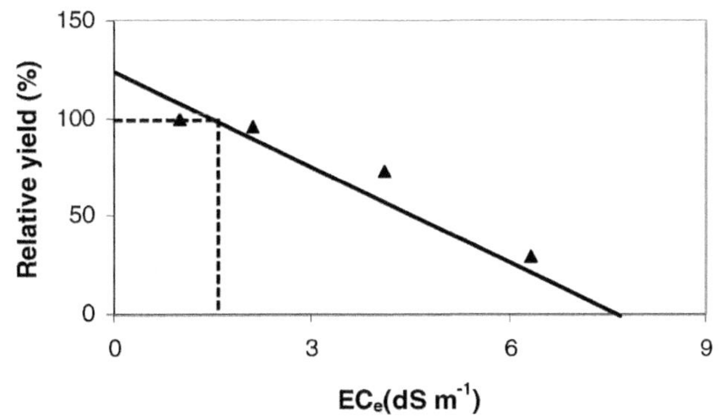


**Figure 4.** Relative yield response of drip irrigated cucumber to soil salinity

The experimental results showed that the $EC_i$ threshold value was bigger than the $EC_e$ threshold. This result is in agreement with the findings of Bahceci [13]. In similar soil and environmental conditions in Konya-Turkey, Bahceci [13] reported that $EC_i$(4 dS $m^{-1}$) threshold value of sprinkler irrigated bean was higher than that of $EC_e$(0.81 dS $m^{-1}$). The researcher explained this situation with low initial soil salinity. In our experiment, it was most probably caused by both low initial soil salinity and the chosen irrigation method itself.

## Conclusions

The use of drip irrigation systems in semi-arid regions shows a great potential to use the low quality irrigation water in irrigation with necessary precautions. Although the experiment period was only 1 year, the results obtained provided useful insights into the effects of salinity on yield, growth and ET. If the soil is not saline, irrigation water salinity up to 3 dS $m^{-1}$ would not significantly affect the yield of drip irrigated cucumber. However, irrigations with low quality water above 3 dS $m^{-1}$ reduced the yield significantly. In particular, the negative effect of the $SW_3$ treatment was much more severe compared to other saline treatments.

## References

1 Katerji N, van Hoorn JW, Hamdy A, Mastrorilli M, Karam F (1998) Salinity and drought, a comparison of their effects on the relationship between yield and evapotranspiration. *Agricultural Water Management* 36: 45–54

2 Stewart JI, Danielson RE, Hanks RJ, Jackson EB, Hagan RM, Pruitt WO, Franklin W, Riley JP (1977) Optimizing crop production through control of water and salinity levels in the soil. Utah Water Research Lab. PR. 151-1, Logan, Utah, 191

3 Shalhevet J (1994) Using water of marginal quality for crop production: Major issues. *Agricultural Water Management* 25: 233–269

4 Shalhevet J, Hsiao TC (1986) Salinity and drought: A comparison of their effects on osmotic adjustment, assimilation, transpiration and growth. *Irrig Sci* 7: 249–264

5 Yamasaki S, Dillenburg LR (1999) Measurements of leaf relative water content in araucaria angustifolia. *Revista Brasilleira de Fisiologia Vegetal* 11(2): 69–75

6 Maas EV, Hoffman GJ (1977) Crop salt tolerance-current. *ASCE Journal of Irrigation and Drainage Division* 103: 115–134

7 Sonneveld C, Voogt SJ (1978) Effects of saline water on glasshouse cucumber. *Plant and soil* 49: 595–606

8 Vulkan-Levy R, Ravina I, Mantell A, Frenkel H (1998) Effect of water supply and salinity on pima cotton. *Agricultural Water Management* 37: 121–132

9 Katerji N, van Hoorn JW, Hamdy A, Mastrorilli M (2004) Comparison of corn yield response to plant water stress caused by salinity and by drought. *Agricultural Water Management* 65: 95–101

10 Maggio A, Pascale SD, Angelino G, Ruggiero C, Barbieri G (2004) Physiological response of tomato to saline irrigation in long-term salinized soils. *Eur J Agronomy* 21: 149–159

11 Ayers RS, Westcot DW (1985) Water quality for agriculture. FAO Irrigation and drainage paper 29, Rome, 174

12 Chartzoulakis KS (1992) Effects of NaCI salinity on germination, growth and yield of greenhouse cucumber. *J Horticultural Sci* 67(1): 115–119

13 Bahceci I (1995) Evaluation of salinity-yield-applied irrigation water interactions for field bean under sprinkler irrigation method. Ph.D. dissertation, University of Cukurova, 109

14 Allen RG, Pereira LS, Raes D, Smith M (1998) Crop evapotranspiration. Guidelines for computing crop water requirements. FAO irrigation and drainage paper No 56, p301, FAO-Rome

Biosaline Agriculture and Salinity Tolerance in Plants
Edited by M. Öztürk, Y. Waisel, M.A. Khan and G. Görk


# Potential utilisation of halophytes for the rehabilitation and valorisation of salt-affected areas in Tunisia

Chedly Abdelly[1], Zouhaier Barhoumi[1], Tahar Ghnaya[1], Ahmed Debez[1], Karim Ben Hamed[1], Riadh Ksouri[1], Ons Talbi[1], Fethia Zribi[1], Zeineb Ouerghi[1], Abderrazzak Smaoui[2], Bernhard Huchzermeyer[3] and Claude Grignon[4]

[1] *Adaptation des Plantes aux Stress Abiotiques, INRST, BP 95, Hammam-Lif 2050, Tunisia*
[2] *Caractérisation et Qualité de l'Huile d'Olive, INRST, BP 95, Hammam-Lif 2050, Tunisia*
[3] *Institut für Botanik, Universität Hannover, Hannover, Germany*
[4] *Biochimie and Physiologie Moléculaire des Plantes, Agro-M INRA, Montpellier, France*

## Introduction

In arid and semi-arid regions, irrigation water contributes to salinisation of the upper layer of the soil, where most root activity takes place. Along the path of plant domestication, many crop species have lost resistance mechanisms to various stress conditions [1], including salt stress [2]. Thus, most crop plants do not fully express their original genetic potential for growth, development and yield under salt stress, and their economic value declines as salinity levels increase [3, 4]. Improving salt resistance of crop plants is, therefore, of major concern in agricultural research. A potential genetic resource for the improvement of salt resistance in crop plants resides among wild populations of halophytes [5, 6]. These can be either domesticated into new, salt-resistant crops, or used as a source of genes to be introduced into crop species by classical breeding or molecular methods.

Given the progressive scarcity of freshwater resources and soil salinisation, a major aim of investigations is to evaluate the potential of halophytic species to be widely and economically used in arid and semi-arid regions. It would encourage the sustainable use of halophytes for the creation of productive ecosystems and re-greening degraded areas, by building up a collection of halophytes with a high tolerance to salt stress and characteristics potentially exploitable from an economic point of view. Among the known 2,600 halophytic species, some present economic (human feeding, fodder, materials of high economic values) or ecological interests (soil desalinisation, dune fixation, phytoremediation, landscaping and ornamentation).

Within the framework of this approach, the Laboratory of Plant Adaptation to the Abiotic Stresses, in the National Institute of Scientific and Technical Research (INRST) of Tunisia has initiated an exploration and a physiological and biochemical characterisation of some halophyte species in order to identify the most promising ones. There are two major topics: the implication of halophyte species in the improvement of soil characteristics (desalinisation and fertilisation, heavy metal extraction), and their economical interests as oleaginous and fodder crops, for instance.

## Results

*Ecological interests of halophytes*

*Improvement of soil characteristics*
Vegetation in saline habitat such as sebkha is heterogeneous. Numerous perennial tufts of strict halophytes are associated with annual species sensitive to salt and mineral deficiency stresses. *Medicago*, characterised by a high fodder value, largely contributes to the ecosystem primary production in the absence of water constraints. These annuals mainly develop within or very close to halophyte tufts. Parallel field and laboratory studies have shown that *Medicago* is sensitive to salinity [7, 8], as well as to nitrogen and phosphorus deficiencies [9]. Furthermore, the shoots of the annuals growing in association with halophyte species contains relatively low $Na^+$

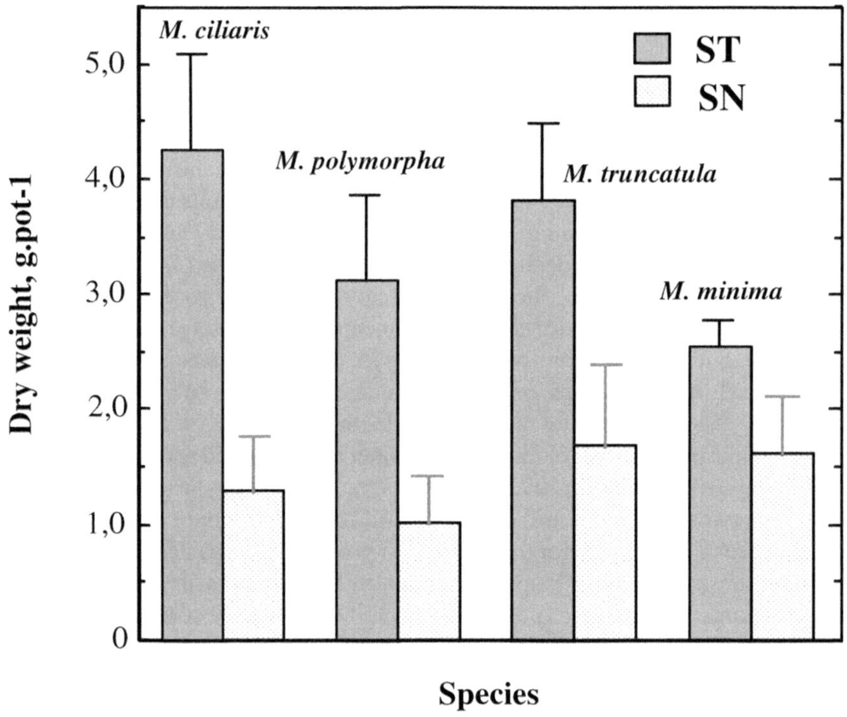


**Figure 1.** Changes in biomass production of *Medicago ciliaris, M. polymopha, M. truncatula* and *M. minima* (g.pot$^{-1}$) with soil origin. ST: culture carried out on soil sampled under halophytes tufts, SN: cultures carried out on soil sampled in outside of halophytes tufts. Means of 20 repetitions and confidence intervals at 95 %

concentrations. These data suggest that the upper horizon of spots near the halophyte tufts (where sensitive annuals grew), is fertile and contains low salt levels. Indeed, this was confirmed by the study of soil samples taken from the upper horizon in the tuft centres. Desalinisation of the upper horizon by the superficial roots of halophytes could be responsible for this microgradient of salinity. Moreover, the litter formed by halophyte fallen organs and by organic debris accumulated by the wind at the vicinity of halophyte tufts, could contribute to localised soil enrichment in N and P. This was confirmed using a biological test of soil fertility. Some tufts of *Salicornia arabica* were removed for sampling soil in the upper horizon (0–20 cm), where roots of annual plants developed. Other samples were taken between halophytes tufts, in zones devoid of vegetation or weakly populated. Four annual *Medicago* species (*M. ciliaris*, *M. polymorpha*, *M. truncatula* and *M. minima)* were grown on these soil samples, without mineral fertilisation, in a greenhouse under controlled conditions. The plants were harvested at the flowering stage. In the four species, total biomass production (dry matter per pot) was higher on soils sampled under the halophyte tufts than on soils from nude zones (Fig. 1). These studies show that perennial halophytes improve soil characteristics by lowering its salt content and by increasing nitrogen and phosphorus concentrations.

The capacity of desalination of saline soil by halophytes was also evaluated in strictly controlled conditions, using *Sesuvium portulacastrum*, an Aizoaceae. After clonal multiplication, the plants were cultivated for 2 months on saline soil, originating from the edge of a sebkha. They were irrigated with a nutrient solution deprived of $Na^{+}$and $Cl^{-}$, without losses by drainage. Salt export by plants was evaluated by the difference between the quantities of $Na^{+}$ and $Cl^{-}$, initially measured in the culture

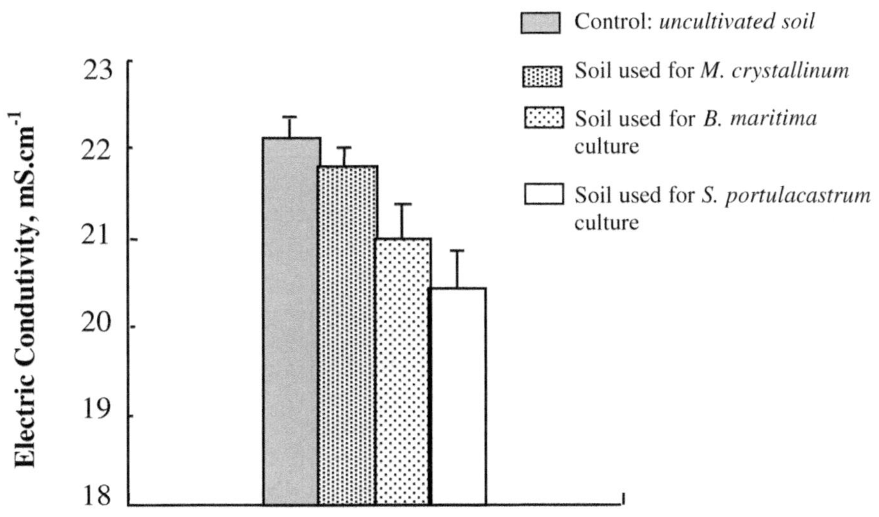


**Figure 2.** Electric conductivity ($mS.cm^{-1}$) of the aqueous extract (1/10) of soil used during two months for halophytes culture. Means of 12 replicates. Bars indicate ± standard errors (p = 0.05)

substrate, and those found at the end of the experiment in the soil. *S. portulacastrum* produced more biomass than other species used in the same experiment (*Batis maritima* and *Mesembryanthemum crystallinum*), and accumulated larger amounts of $Na^+$ and $Cl^-$(about 6.5 mmol.g$^{-1}$ DW, amounting to 30–40 % of the biomass). At the end of the cultures, the soil used for the culture of *S. portulacastrum* showed a significant (10%) decrease of its salinity estimated by electric conductivity (Fig. 2). This study demonstrates that the associated characteristics of *S. portulacastrum*, namely high growth rate and high capacity for salt accumulation, permit soil desalination, even in short-term cultures. Thus, this species would be interesting for the rehabilitation of the saline lands.

*Heavy metal extraction*

In Tunisia, saline depressions, colonised by halophyte species, often constitute sites of accumulation of industrial effluents contaminated by heavy metals. Indeed, preliminary studies achieved in various regions of Tunisia showed that these zones are contaminated by cadmium, nickel and lead. We studied the response to Cd of two halophyte species, *S. portulacastrum* and *M. crystallinum*. In the absence of Cd, the biomass of *M. crystallinum* plants was much larger than that of *S. portulacastrum*. However, Cd severely inhibited *M. crystallinum* growth, even at the lowest concentration (50 µM), but did not significantly modify that of *S. portulacastrum*. In the shoots, the Cd concentrations in *S. portulacastrum* shoots was half (100–350 ppm)

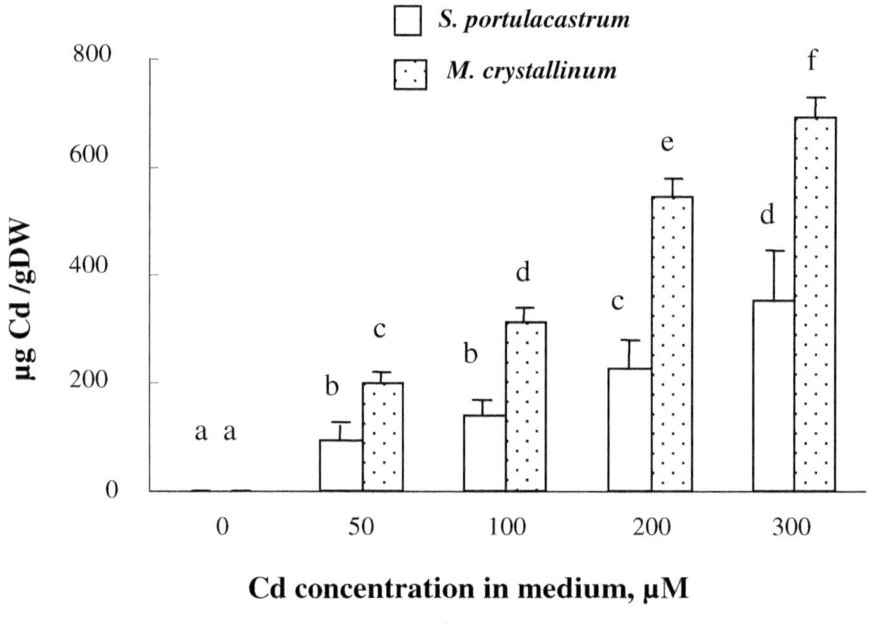


**Figure 3.** Changes in Cd concentration (µg.g$^{-1}$ DW) in shoots of *S. portulacastrum* and *M. crystallinum* treated by various Cd concentrations. Means of 8 replicates. Bars marked with same letter are not significantly different at $p = 0.05$

that in *M. crystallinum* shoots (200–700 ppm) (Fig. 3). According to these data, both species would be classified among Cd hyper accumulator plants and would be of this fact interesting for the phytoremediation. The analysis of the relationship between growth and mineral status in the two halophytes suggested that the Cd-induced decrease of growth resulted not from direct effect of accumulated Cd, but rather from restriction of $K^{+}$, $Ca^{2+}$, and $Fe^{2+}$ uptake. This hypothesis was studied using a split-root system: after a pretreatment phase, seedlings were divided into three lots. Half of the roots of the first lot were immersed in Cd free medium, while the other half were immersed in the same medium supplemented with 100 μM Cd (B/Cd). For the two other lots, the two halves of the root system were immersed either in free Cd medium (B/B) or in medium supplied with 100 μM Cd. In comparison with Cd/Cd plants, the split root Cd/B plants displayed improved growth. This effect was associated to an increase in nutrient uptake. Furthermore, the Cd/B plants accumulated Cd at a level similar to that of Cd/Cd plants. In summary, our results indicate that the Cd-induced decline growth resulted rather from an indirect Cd effect (inhibition of nutrient uptake) than from a direct Cd effect (excess of Cd accumulation) and suggest the possibility to increase the capacity of the two halophyte species to extract Cd while improving nutrient availability in the medium.

## *Economical interests of halophytes*

### *Halophytes with fodder potential*

In Tunisia, fodder crops occupy currently only 7 % of the cultivated zones. In addition, the surface of the uncultivated saline area is four times more important than that of the pastures. The identification of fodder halophytes would make it possible to exploit new zones of production and to reduce our deficit in fodder. Within the framework of this approach, several potentially suitable species were identified. The conditions of their multiplication and culture were established as well as their salt tolerance limits and their nutritional requirements.

*Suaeda fruticosa*, an indigenous Chenopodiaceae in Tunisia, is quite frequent in semi-arid, arid and desertic bioclimatic stages and well appreciated by livestock [10]. *Spartina alterniflora*, a Poaceae, is dominant in saline marsh and coastal regions in the east of USA. For *S. fruticosa*, maximal dry matter production occurred at NaCl concentrations comprised between 100–300 mM, with a 85-fold increase in dry weight following a 45 day treatment. *S. alterniflora* expressed maximal growth when irrigated with nutrient solution or containing 0–100 mM NaCl, its initial dry weight being six-fold increased after 100 days of treatment. *S. fruticosa* was more salt tolerant than S. *alterniflora* under moderate NaCl concentrations (300 mM NaCl), but more sensitive at the highest NaCl concentration (800 mM). Considering the decreasing availability of freshwater in arid regions, the utilisation of non-conventional water resources (brackish water, waste water, and seawater) constitutes a promising approach, especially as these halophytes require salt to express their maximal growth potentialities. However, the growth of both species was limited when they were irrigated with seawater, owing to the low availability of some nutrients. Indeed, they displayed higher growth rates when nitrogen and, to a lesser degree, phosphorus were added to the seawater [11].

The capacity of biomass production is an important characteristic which must be considered in the evaluation of the fodder halophytes. Using cultures carried out in pots of 0.85 $m^3$, we demonstrated that *S. alterniflora* could produce 7,500 kg DW per hectare in one cut. As the season of growth of this species extends from March–October, it is possible to carry out at least two cuts, which will ensure a primary production of 15 tons DW per hectare and per year. Similar estimation showed that *S. fruticosa* could produce at least 4,500 kg DW per hectare. These yields are comparable to those of the conventional fodder crops irrigated with freshwater, like alfalfa, 10 tons.$ha^{-1}$ per year [12], and clover, 8 tons.$ha^{-1}$ per year [13]. Analysis of published data indicates that the yield of the most productive halophytes varies from 8–17 tons DW.$ha^{-1}$ per year. The comparison of the mineral composition of the shoots of both halophytes with that of the fodder required by livestock showed that excepting $Ca^{2+}$ content in *S. alterniflora*, $K^+$, $Mg^{2+}$ and P concentrations in tissues exceed the nutritional requirements of the cattle. The content of total nitrogen is higher in *S. fruticosa* than in *S. alterniflora*, but the digestible nitrogen fraction is similar in both halophytes, and meets perfectly the nutritional requirements of the sheep. According to Glenn and O'Leary [14] and Bayoumi et al. [15], proteins represent 15 % of DW in several halophytes. According to these data, *S. fruticosa* and *S. alterniflora* can annually produce 1.5 tons of proteins for an average biomass of 10 tons.$ha^{-1}$. The quantitative and qualitative yield of the two halophytes is appreciably similar to that of alfalfa irrigated with freshwater [16].

Two other halophytes from Tunisia, with fodder potential, were characterised: *Aeluropus littoralis* and *Catapodium rigidum*, respectively perennial and annual Poaceae. In the absence of salt, *Catapodium* displayed a relative growth rate (RGR) slightly higher than that of *Aeluropus*. Salt decreased this parameter in both species. On RGR basis (expressed as % of control without salt) *A. littoralis* was more tolerant than *C. rigidum*, this behaviour being accentuated with increasing salinity. In salinity range not exceeding 400 mM, RGR remained between 0.04–0.06 $day^{-1}$, values characteristic of spontaneous or cultivated *Medicago* species not subjected to salt, 0.08–0.09 [7, 9] or others fodder halophytes *Suaeda fruticosa*, 0.07–0.09 [11], *S. alterniflora*, 0.03 [11], *Spartina anglica*, *Puccinellia maritima*, 0.02–0.05 $day^{-1}$ [17].

Both halophytes accumulated sodium mainly in their shoots. However, $Na^+$ concentrations were lower in *A. littoralis* than in *C. rigidum* (maximum values around 5 mmol. $g^{-1}$ DW and 10 mmol. $g^{-1}$ DW, respectively). In addition to its capacity to control $Na^+$ transport towards shoots, *A. littoralis* secreted more than 50% of leaf $Na^+$ by salt glands. Indeed, $Na^+$ accumulated inside tissues did not exceed 2 mmol.$g^{-1}$ DW under the most severe salinity. This secretion seems to be selective for $Na^+$ and $Cl^-$, since $K^+$ was completely absent among the elements secreted on the surface of the leaves.

*Oleaginous halophytes*

Some Tunisian salty areas have been explored for plants which could be considered as oilseed species. Three potentially interesting species have been identified as oil producers, *Zygophyllum album* (Zygophyllaceae), *Cakile maritima* (Brassicaceae), and *Crithmum maritimum* (Apiaceae). Oil extraction was carried out on ripe seeds collected in their natural biotope. Some physiological and biochemical aspects were studied, such as individual mass of seeds, oil content, and lipid and fatty acid composition. The seeds of investigated species have a suitable size for harvesting. The dry weight of 100 seeds ranged from 133 mg in *Zygophyllum album* to 774 mg in *Cakile maritima*. The value for the latter species is nearly three times higher that for rape (*Brassica napus*) seeds, a conventional oleaginous plant. The seeds of the two other halophytes are smaller than *Cakile maritima* seeds, with mass approximately half of that of rape ones.

The seeds of *Cakile maritima* and *Crithmum maritimum* present high levels of oil, reaching respectively 42 % and 30 % of the seed DW. Oil content in *Zygophyllum album* seeds is very low (6 %). As for olive oil, fatty acid composition of *Crithmum maritimum* seeds is characterised by a high level of oleic acid (81 %), whereas that of *Zygophyllum album*, characterised by a high percentage of linoleic acid (64 %), is similar to sunflower oil composition. Therefore, these two species contain oils of good quality which can be used without any further modification. Oil of *Cakile maritima*, rich in erucic acid (25 %) may be used for industrial applications [18].

In the laboratory, growth of *Cakile maritima* was stimulated under moderate (50–100 mM) concentrations of NaCl. The response of the whole plant was essentially due to the salt effects on shoots, their growth being significantly augmented at 100 mM NaCl. In the 200–300 mM range, the whole plant biomass production was maintained at approximately 90 % of the control. At higher NaCl concentrations

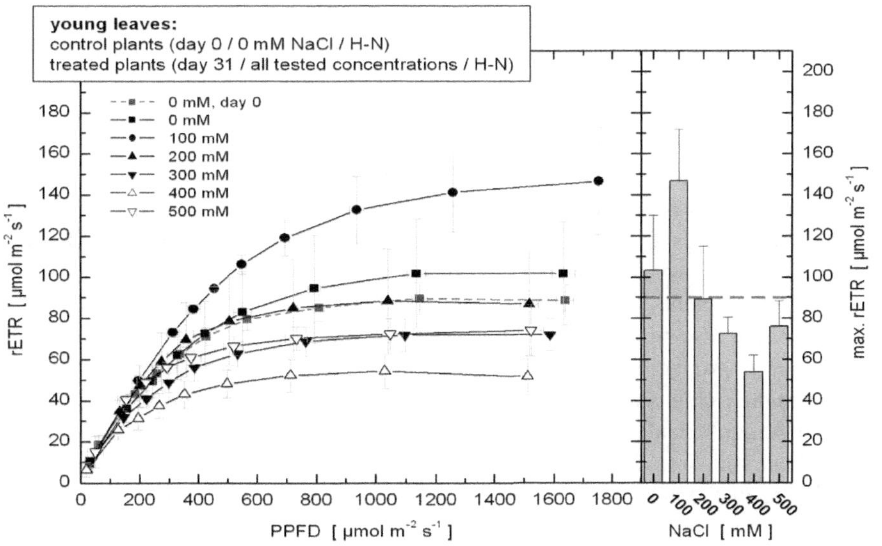


**Figure 4.** NaCl effect on photosynthesis. Changes in the electron transport rate (ETR) in the leaves of *Cakile maritima* subjected to increasing NaCl concentrations for 35 days. Means of 10 replicates ± standard errors (p = 0.05)

(500 mM), an important and significant reduction of growth was observed, but plants survived [19]. Concerning photosynthesis, various parameters were measured: gas exchange, net assimilation of $CO_2$, and electron transport rate (Fig. 4). As growth rate, these parameters indicated that the optimal physiological functioning was obtained with 50–100 mM NaCl. In these optimal conditions, the number of seeds per plant was significantly augmented by salt. At higher concentrations, a significant reduction of seed production was observed. All salt treatments resulted in seeds significantly smaller than in control. Culture of *Cakile maritima* in the presence of salt modified the biochemical composition of seeds: the oil content and the rate of erucic acid, largely used in industry, were augmented.

## Conclusions

These studies were aimed at identifying among the halophytes promising species to ensure a plant productivity of economic and/or ecological interest, in the marginal zones and under conditions of irrigation with non-conventional water resources (brackish water, waste water and even seawater, more or less diluted). We identified several halophytes interesting for livestock nutrition. Monocotyledonous with a salinity avoidance strategy, enabling them to produce a biomass containing relatively little salt (*Spartina alterniflora*, *Aeluropus littoralis*, *Catapodium rigidum)* are particularly interesting. Culture conditions of these plants, their limits of salt tolerance and their nutritional requirements, are known. The data obtained so far are promising. Work is in progress to characterise their response to other constraints which could limit their yield in saline lands and particularly the low nitrogen availability [20]. Indeed, nitrogen is the most limiting nutrient for plant growth in saline ecosystems. Since a greater availability of $NH_4^+$ than $NO_3^-$ was often observed, the research of the halophytes able to use ammoniacal nitrogen would be an effective and less expensive substitute than the use of nitrogen mineral fertilisers.

Two promising oleaginous halophytes were identified: *Cakile maritima* and *Crithmum maritimum*. An important intraspecific variability was observed within these two species at the levels of i) seed biochemical characteristics (oil content, fatty acid composition), and ii) the physiological response to salt. Studies carried out on these plants from germination to the seed maturation showed that *Cakile maritima*, producing industrial oil, is relatively more tolerant to salt than *Crithmum maritimum*, producing edible oil [21]. However, recent data showed that *Crithmum maritimum*, rich in antioxidant molecules, is also interesting for medicinal purposes.

At the ecological level, particular interest was paid to the halophytes able to fix and desalinate soil. *Sesuvium portulacastrum* with its growth stimulated by high (800 mM salt) concentration under adequate mineral nutrition is a promising candidate. The analysis of its responses to heavy metals also suggests that this species is interesting for phytoremediation.

*Acknowledgements*
This work was supported by the following projects: Concerted Action: IC18CT96-0055 *Sustainable Halophyte Utilisation in the Mediterranean and Subtropical Dry Regions, 1996–2000*; Program of research supported by the Secretariat of State to Scientific Research and Technology *Étude d'espèces halophytiques d'intérêts économiques et écologiques, 1998–2002*; Network CMCU02/F0924 supported by the French-Tunisian co-operation, *Mise au point d'outils physiologiques et moléculaires d'identification des populations locales de halophytes pour valoriser les sols salins, 2002–2004*; and a Program of research supported by the Ministry for Scientific Research, Technology, and Development of Competences *Utilisation des halophytes pour la réhabilitation et la valorisation des sols salins, 2002–2005.*

## References

1 Serrano R (1996) Salt tolerance in plants and microorganisms: toxicity targets and defence responses. *Int Rev Cytol* 165: 1–52
2 Munns R (1993) Physiological processes limiting plant growth in saline soils: some dogmas and hypotheses. *Plant Cell and Environ* 16: 15–24
3 Läuchli A, Epstein E (1990) Plant responses to saline and sodic conditions. In: KK Tanji (ed.) In: *Agricultural Salinity Assessment and Management,* ASCE manual No. 71, New York, 113–137
4 Maas EV (1990) Crop salt tolerance. In: KK Tanji (ed.) In: *Agricultural Salinity Assessment and Management,* ASCE manual No. 71, New York, 262–304
5 Glenn E, Brown JJ, Blumwald E (1999) Salt tolerance and crop potential of halophytes. *Crit Rev Plant Sci* 18: 277–255
6 Serrano R, Mulet J, Rios G, Marquez J, de Larrinoa I, Leube M, Mendizabal I, Pascual-Ahuir A, Proft M, Ros R et al. (1999) A glimpse of the mechanisms of ion homeostasis during salt stress. *J Exp Bot* 50: 1023–1036
7 Abdelly C, Lachâal M, Grignon C, Soltani A, Hajji M (1995) Association épisodique d'halophytes stricts et de glycophytes dans un écosdystème hydromorphe salé en zone semi-aride. *Agronomie* 15: 557–568
8 Abdelly C, Zid E, Hajji M, Grignon C (1996) Biomass production and nutrition of *Medicago* species associated to halophytes on the edge of a sebkha in Tunisia. In: R Choukr-Allah, CV Malcolm, A Hamdy (eds) In: *Halophytes and Biosaline Agriculture.* Dekker Inc., New York, 313–324
9 Abdelly C (1997) Mécanismes d'une association de Luzernes spontanées et de halophytes pérennes en bordure de sebkha. Thèse de Doct., Faculty of Sciences of Tunis, Tunisia, 290
10 Le Houérou HN (1996) Forage halophytes in the Mediterranean basin. In: R Choukr-Allah, CV Malcolm, A Hamdy (eds) In: *Halophytes and Biosaline Agriculture.* Dekker Inc., New York, 115–136
11 Sleimi N, Abdelly C, (2002) Growth and mineral nutrition of some halophytes under seawater irrigation. In: R Ahmad, Malik KA (eds) In: *Prospects for Saline Agriculture,* 403–410
12 Martiniello P (1998) Influence of agronomic factors on the relationship between forage production and seed yield in perennial forage grasses and legumes in a Mediterranean environment. *Agronomie* 18: 591–601
13 Bublot G, Hallet A (1981) Aspects économiques de la production de fourrages grossiers. I. Economie des productions fourragères: principe et évolution. *Rev Agric* 3: 421–433

14 Glenn EP, O'Leary JW (1984) Relationship between salt accumulation and water content of dicotyledonous halophytes. *Plant Cell Environ* 7: 253–261
15 Bayoumi MT, El Shaer HM, Assaad F (1990) Survival of sheep and goats fed with salt march plants. *J Arid Environ* 18: 75–78
16 Glenn E, Hicks N, Riley J, Swingle S (1996) Seawater irrigation of halophytes for animal feed. In: R Choukr-Allah, CV Malcolm, A Hamdy (eds) In: *Halophytes and Biosaline Agriculture*. Dekker Inc., New York, 221–236
17 Rozema J, Van Dieggelen J (1991) A comparative study of growth and photosynthesis of four halophytes in response to salinity. *Acta Œcol* 12: 673–681
18 Zarrouk M, El Almi H, Ben Youssef N, Sleimi N, Ben Miled D, Smaoui A, Abdelly C (2003) Lipid composition of seeds of local halophytes secies: *Cakile maritima*, *Zygophyllum album* and *Crithmum maritimum*. In: H Lieth (ed.) In: *Cash Crop Halophytes: Recent Studies*. Kluwer Academic Publishers Group, 121–126
19 Debez A, Ben Hamed K, Grignon C, Abdelly C (2004) Salinity effects on germination, growth and seed production of the halophyte *Cakile maritima*. *Plant Soil* 262: 179–189
20 Messedi D, Labidi N, Grignon C, Abdelly C (2004) Limits imposed by salt to the growth of the halophyte *Sesuvium portulacastrum*. *J Plant Nut Soil Sci* 167: 720–725
21 Ben Hamed K, Debez A, Chibani F, Abdelly C (2004) Salt response of *Crithmum maritimum*, an oleaginous halophyte. *J Trop Ecol* 1: 151–159

Biosaline Agriculture and Salinity Tolerance in Plants
Edited by M. Öztürk, Y. Waisel, M.A. Khan and G. Görk


# Interactive effect of potash and organic manures on growth and nutrient uptake of sugarcane grown under saline conditions

Muhammad Y. Ashraf[1], Faqir Hussain[1], Muhammad M. Iqbal[1], Wolfgang Maibaum[2] and Marcus Ross[2]

[1] *Stress Physiology and Plant Biochemistry Lab, Nuclear Institute for Agriculture and Biology (NIAB), PO Box 128, Jhang Road, Faisalabad, Pakistan*
[2] *Agricultural Advisory Department, K+S Kali GmbH, Kassel, Germany*

## Salinity and crop production

Soil salinity poses severe problems to crop production in many countries and this problem is very severe in arid and semi-arid regions of the world which occupy one third of the area of the earth (4.3 billion ha). Salinity is among the serious problems of irrigated agriculture of Pakistan. Millions of hectares of potentially productive land is uncultivatable due to excess of salt. The population of Pakistan is increasing at 2.61 % and the gap between the supply and demand of agricultural products is widening day by day. In order to meet the future demand of food, fiber, fuel and industrial raw material, the extension of agriculture would require the use of marginal lands [1]. Most of Pakistan is arid to semi-arid and has a low annual precipitation. Of the 20.36 mha of the total cultivated land, 6.3 mha are salt-affected [1]. A major part of the salt affected soils (about 3.5 mha) are presently cultivated to rice, wheat, cotton and sugarcane, but the output is very low. According to an estimate, the reductions in the yield of rice, wheat, cotton, sugarcane cultivated on such moderately salt-affected soils are, 64, 62, 59 and 68 %, respectively [2].

## Role of potassium in plant growth and metabolism

Potassium is an essential nutrient element for all plants and in most terrestrial plants $K^+$ is the major cationic inorganic nutrient element. Potassium acts to balance the charge in the cytoplasm of plant cells, where $K^+$ is the dominant counter ion for the large excess of negative charge on proteins and nucleic acids [3]. It activates the crucial enzymatic reactions such as those occurring in the formation of pyruvate and is also a substantial contributor to the osmotic pressure of the vacuole and hence to cell turgor which endows non-lignified plant cells with structural rigidity. In contrast $Na^+$ is only required for halophytes (for translocation of pyruvate across the chloroplast envelope) where it acts as a micronutrient [4]. In most other species $Na^+$ does not act as a nutrient in the sense that it is strictly required for growth but its addition to the growth medium may promote growth of many plants when the $K^+$ supply is limited and particularly the growth of salt tolerant plants, by contributing to turgor formation [5].

## Soil productivity and organic/inorganic amendments

A major agricultural research priority is to sustain soil productivity and to develop better methods to monitor changes in soil physical, chemical and biological properties as affected by soil management. It is well established that organic addition of amendments to soil can positively affect soil as well as crop productivity. An organic manure amendment to soil increases soil fertility and porosity, microbial biomass and microbial activity [6]. There are many reports, which indicate that application of organic matter increases soil urease and phosphatase activities and add N and organic matter to soil. Microbial degradation and mineralization of organic matter provides nutrients to become available to the growing crop [7]. The amendment of organic matter also increases the moisture holding capacity and the $Ca^{2+}$ exchange capacity and decreases soil pH as a result of which micronutrients become available to plants [6].

Heavy yield losses were observed in different crops due to salinity. According to an estimate, the yield losses in sugarcane on moderately salt-affected areas were up to 62 % [8]. Different approaches are being used to utilize or to reclaim the salt-affected soils. Presently, we have planned to utilize the moderately salt-affected soils for sugarcane production by the management of Na/K ratio. Potassium is an essential major element involved in maintaining the water status of plant and turgor pressure of its cells and opening and closing of its stomata. Potassium is required in the accumulation and translocation of newly formed carbohydrates.

Keeping in mind the importance of organic matter and potassium for soil and plant productivity, experiments on four sites (one normal and three salt affected) of Punjab, Pakistan, are being conducted to investigate the relative response of sugarcane crop to SOP ($K_2SO_4$) versus MOP (KCl) with amendment of two types of organic manures in salt-affected soils.

## Field studies and soil/plant analysis

The interactive effect of Potash, i.e., SOP or MOP, and two organic fertilizers, i.e., farmyard manure and sugarcane press mud, on the yield and quality of two sugarcane varieties, SPSG-26 (salt tolerant) and CP-77-400 (salt sensitive), was studied at three selected sites in Pakistan which are normal, saline and saline-sodic. The soil and water samples of three selected sites, i.e., NIAB Faisalabad, Jhang, and Samundri, have been collected and analyzed for various physicochemical properties, i.e., soil texture was determined by hydrometer method [9], water holding capacity estimated by Hillguard method, bulk density by core sampler technique (100 $cm^3$), permanent wilting

**Table 1.** Soil characteristics of four selected sites to study the effect of SOP and MOP on the yield and quality of sugarcane

| **Soil characteristics** | **Selected sites** | | | |
|---|---|---|---|---|
| | **NIAB** | **JHANG** | **PINDI BHATTIAN** | **SAMUNDRI** |
| Soil Texture | Clay-loam | Sandy-clay-loam | Clay | Clay-loam |
| Ece (dS $m^{-1}$) | 1.03–1.732 | 10.09–19.99 | 4.7–10.8 | 5.26–16.75 |
| PH | 7.4–7.9 | 8.35–8.46 | 8.41–9.29 | 8.81–9.15 |
| SAR | 0.45–0.62 | 27.37–99.2 | 41–78 | 29–86 |
| $Na^+$ (meq $L^{-1}$) | 0.9–1.2 | 75–130 | 80–150 | 60–110 |
| Ca+Mg (meq $L^{-1}$) | 7.5–10.0 | 15.0–20.0 | 7.5–12.5 | 8.5–18.5 |
| $K^+$ (meq $L^{-1}$) | 1–1.25 | 1.2–1.5 | 1–1.25 | 1–1.56 |
| $CO_3{}^{2-}$ (meq $L^{-1}$) | – | – | – | – |
| $HCO_3{}^-$ (meq $L^{-1}$) | – | 25–45 | 25–75 | 28–60 |

**Table 2.** Tubewell waters used for irrigation at four selected sites to study interactive effects of SOP, MOP and organic fertilizers (farmyard manure and sugarcane press mud) on the yield and quality of sugarcane

| **Tubewell water characteristics** | **Selected sites** | | | |
|---|---|---|---|---|
| | **NIAB** | **JHANG** | **PINDI BHATTIAN** | **SAMUNDRI** |
| EC (dS $m^{-1}$) | 0.77 | 2.5 | 2.2 | 1.92 |
| PH | 7.9 | 8.5 | 8.20 | 7.75 |
| SAR | 5.7 | 33 | 24 | 19.12 |
| $Na^+$ (meq $L^{-1}$) | 7 | 25 | 22 | 17.40 |
| $Ca^+$ Mg (meq $L^{-1}$) | 3 | 1.04 | 1.8 | 1.65 |
| $K^+$ (meq $L^{-1}$) | 0.7 | 0.5 | 0.7 | 1.23 |
| $CO_3{}^{2-}$ (meq $L^{-1}$) | – | – | – | – |
| $HCO_3{}^-$ (meq $L^{-1}$) | 2 | 15 | 18 | 12.0 |

point (WP) and field capacity (FC) by Pressure Membrane Apparatus. Electrical conductivity of saturation extract, pH, $HCO_3$, Cl, $SO_4$ and Ca+Mg were determined according to Jackson [10], organic matter according to Walkley and Black method (as described by Jackson [10]), total nitrogen by the Kjeldahl procedure [10]. Sodium and potassium contents were estimated flame-photometerically, whereas phosphorus (available) was extracted by $NaHCO_3$ and determined colorimetrically [11]. All the soil characteristics are summarized in Tables 1 and 2. Experiments on each site consisted of three treatments of organic manure (without manure, farmyard manure and sugarcane press mud at 4 t $ha^{-1}$) in main plots and five treatments of potassium as SOP and MOP (without K, 100, 200 kg $K_2O$ $ha^{-1}$ as SOP and 100 and 200 kg $K_2O$ $ha^{-1}$ as MOP) in subplots with three replications. The seeds of the above-mentioned two sugarcane varieties were obtained from Shakargunj Sugar Mills Ltd, Jhang, Pakistan, and Ayub Agriculture Research Institute Faisalabad, Pakistan. Sowing was done in furrows, and furrow-to-furrow distance was 75 cm at all sites. The sowing was completed during the second and third week of September. The emergence was recorded up to October. Data for plant height was recorded time to time, however, in the present report the data presented on plant height, number of tillers, cane diameter and flag leaf area were recorded during June next year, at all sites. Leaf samples were collected from all four sites, dried and analyzed for $Na^+$, $K^+$, $Ca^{2+}$, P and N.

**Soil and water analyses of selected sites for experimentation**

The soil analyses (Tab. 1) showed that highest soil salinity was recorded at Jhang (10–20 dS $m^{-1}$) followed by Samundri (5–17 dS $m^{-1}$) and NIAB (1–2 dS $m^{-1}$). The pH of all the soils was more than 7 and SAR at Jhang was again maximum ranging from 27–99 followed by Samundri (29–86) and NIAB (less than one). $Na^+$ content in the soil of Jhang was higher than Samundri and NIAB, maximum Ca+Mg content was recorded in the soil samples of Jhang (15–20 meq $l^{-1}$) and minimum at NIAB.

The potassium content of all the soils was almost the same (1–1.56 meq $l^{-1}$). The carbonate and bicarbonates were absent in the soils of NIAB, while the soil samples of other sites contained $HCO_3$. The highest $HCO_3$ were present in soils of Samundri (28–60 meq $l^{-1}$) followed by Jhang (25–45 meq $l^{-1}$). The SAR of Jhang was the highest, which affected plant emergence. Soil pH affected nutrient uptake, especially high pH reduced the availability of nutrients. The texture of Jhang was sandy-clay loam while the other two sites, and NIAB and Samundri were clay-loam.

All the sites were irrigated with tube-well water except NIAB where irrigation was done with canal water. The maximum EC of irrigation water was 2.5 dS $m^{-1}$ at the Jhang site having highest SAR (33) followed by Samundri (1.92 dS $m^{-1}$) with SAR 19 and NIAB (0.77 dS $m^{-1}$) with SAR 5.7 (Tab. 2). The higher EC and SAR are toxic for plant emergence and growth. Similarly greatest Na contents were observed in tube-well water of Jhang (26 meq $L^{-1}$) and minimum in that of NIAB (7 meq $L^{-1}$). The results showed that irrigation water of Jhang is more detrimental than others.

Ca+Mg was highest in the irrigation water of NIAB followed by Samundri and Jhang. High Ca+Mg contents are beneficial for plant growth. Potassium in all irrigation waters is deficient and carbonate is absent in all types of irrigation water used in these experiments. Bicarbonates were recorded in irrigation water and minimum in NIAB. Higher $HCO_3$ are also toxic for plant growth.

## Sugarcane growth and yield

The results showed that salinity reduced sugarcane seed emergence, cane length, diameter and cane yield per plot at all sites under all treatments (Figs. 1 and 2); however, potassium application and amendment with organic manure enhanced seed emergence and all growth and yield parameters. The treatments of SOP or MOP increased growth and yield at all sites with few exceptions. The interactive effect of organic manure and potash showed that SOP performed better than MOP in all organic fertilizers; however, few exceptions were also observed where performance of MOP was better than SOP (Figs. 1 and 2). As the soil salinity in fields was found in patches [12], therefore, the behavior of potash fertilizers was suppressed by high concentrations of salt. However, overall performance of SOP was encouraging in the case of seed emergence, which clearly indicated that SOP was a better fertilizer for sugarcane seed emergence under saline conditions.

The cane yield was calculated on a per hectare basis and the highest yield was obtained from the plants which were growing on normal soil at NIAB followed by saline soil at Samundri, and Jhang (Fig. 1). The application of potash and organic manure significantly enhanced cane yield both under saline and under normal conditions.

Sugarcane variety SPSG-26 performed better than CP-77-400 at NIAB and Samundri while in other places the trend was reversed, which was due to the high salinity in the root media soils selected for SPSG-26. The salt was beyond its tolerant limit, which is why it collapsed. The reduction in cane diameter may be due to the enhancement in cell division and cell division can increase only in those plants which can maintain their turgor potential and from the literature it can be proved that $K^+$ is helpful in maintaining turgor potential in plants [12]. $K^+$ is also involved in many metabolic pathways and had major role in increasing plant growth [13].

*Nutrient uptake*

*$Na^+$ concentration*

$Na^+$ concentration was increased by salinity, lowest $Na^+$ content was observed in the plants growing under normal conditions at NIAB followed by Samundri and Jhang. Application of organic manures also increased the $Na^+$ concentration; however, the highest $Na^+$ concentrations were recorded in those plants that were treated with sugarcane press mud (Fig. 3). The treatments of SOP and MOP reduced the uptake of $Na^+$ at all sites but the SOP was more effective than MOP. The variety SPSG-26 accumulated less amounts of $Na^+$ than CP-77-400 at all selected sites.

*Potassium*

The data clearly indicated that the plants with higher $K^+$ had higher growth and growth parameters and organic manure amendment and potassium application both increased the $K^+$ concentration and its availability for plants (Fig. 3). $K^+$ is involved in maintaining the water status of the plant and turgor potential of its cells. It also has a major role in the opening and closing of stomata. Potassium is also required in the accumulation and translocation of newly prepared carbohydrates [14]. Therefore, the

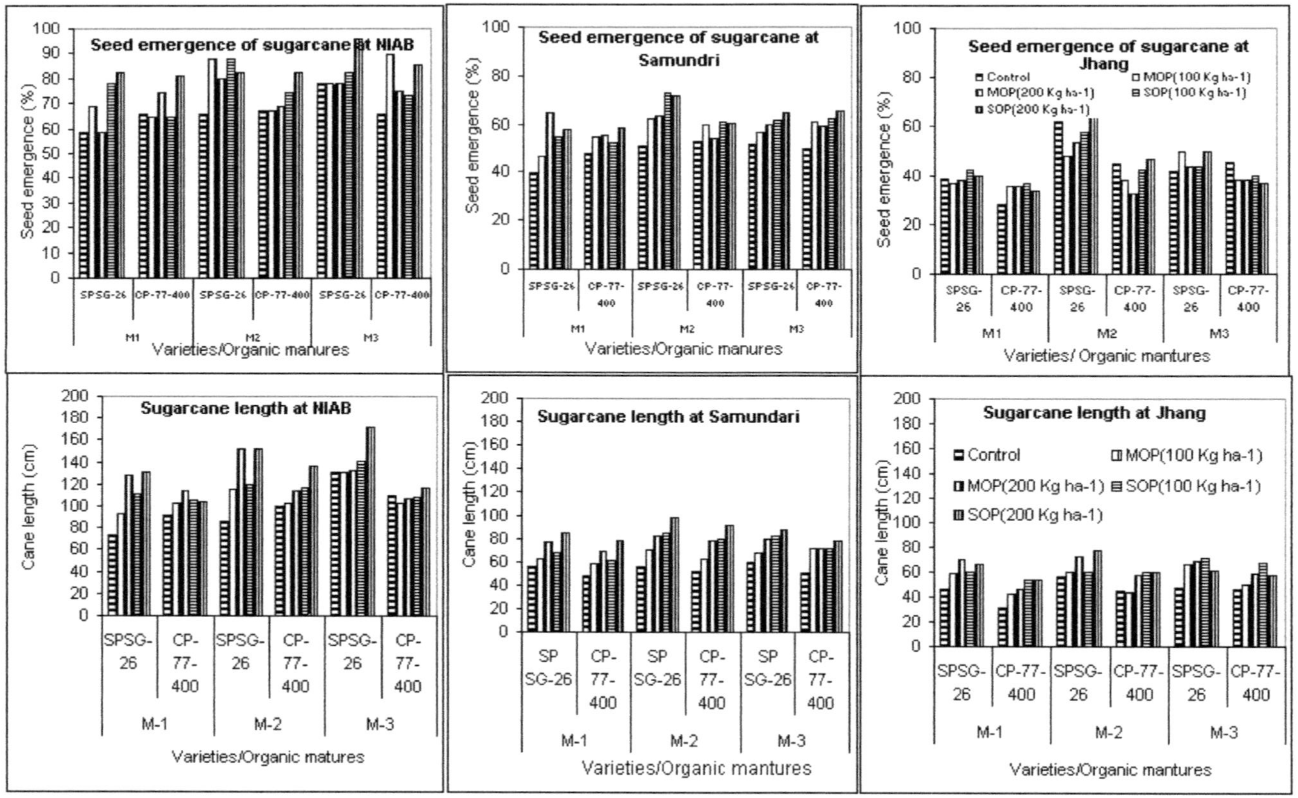


**Figure 1.** Interactive effect of SOP, MOP and organic manures on sugarcane seed emergence and sugarcane at different selected sites under saline and non-saline conditions

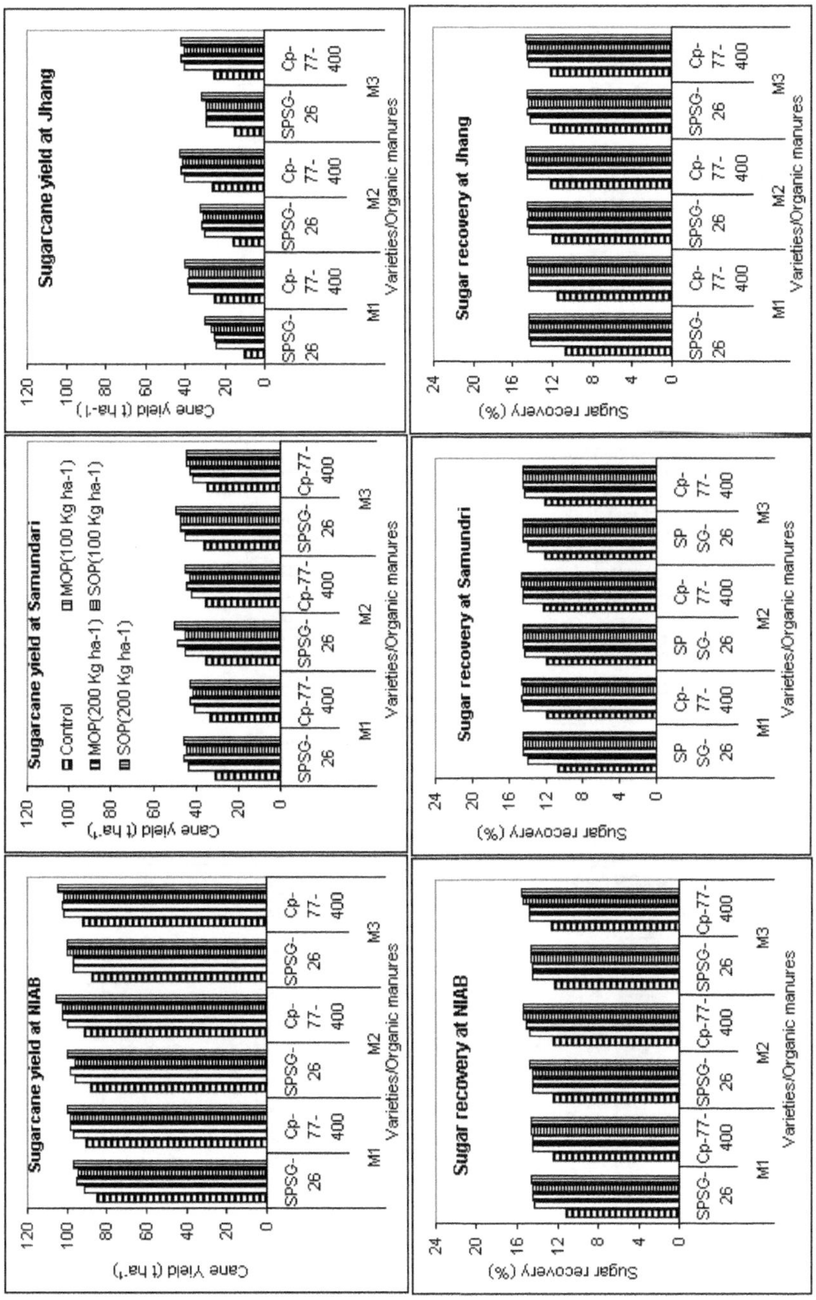


**Figure 2.** Interactive effect of SOP, MOP and organic manures on sugarcane yield and sugar recovery of sugarcane at different selected sites under saline and non-saline conditions

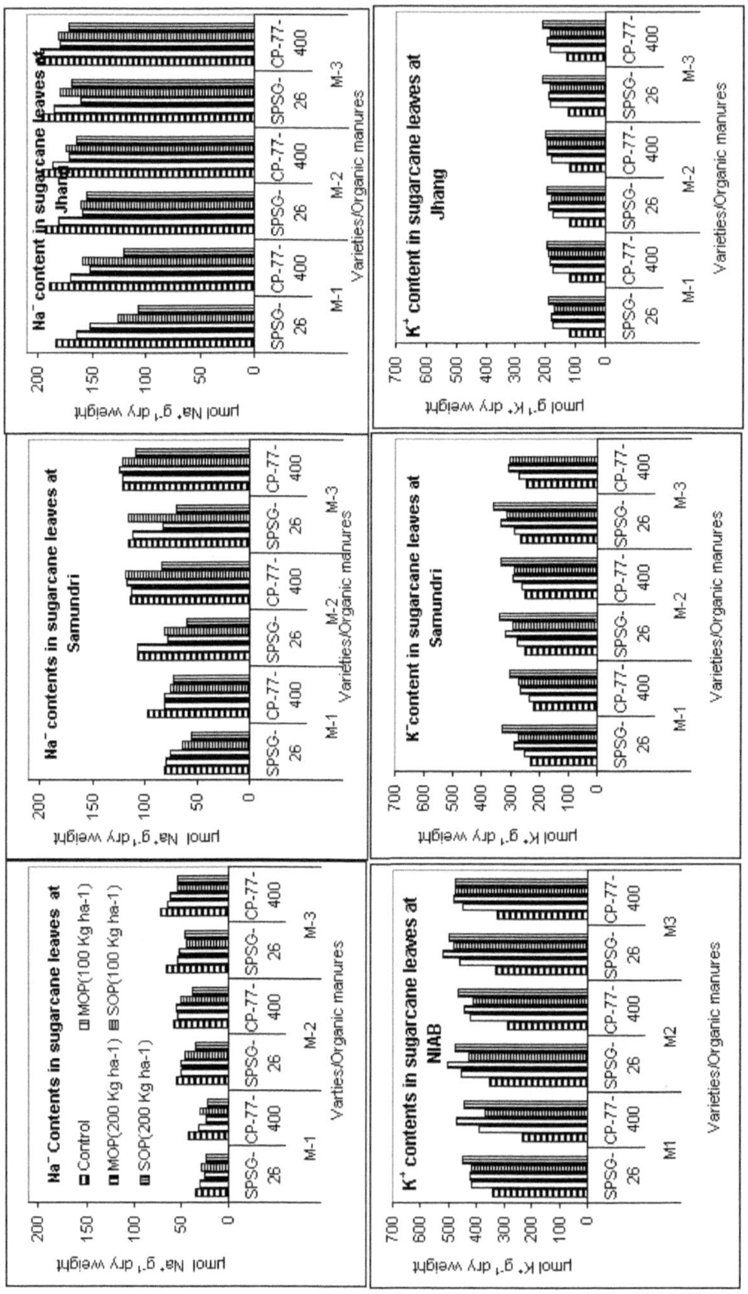


**Figure 3.** Interactive effect of SOP, MOP and organic manures on $Na^+$ and $K^+$ contents in sugarcane leaves at different selected sites under saline and non-saline conditions

better supply of $K^+$ manages the plant growth up to optimum levels and also maintains plants to adjust osmotically with adverse environments [13]. Salt tolerance is not exclusively correlated with adaptations to $Na^+$ toxicity *per se* but also reflects adaptations to secondary effects of salinity, such as water deficit and impaired nutrient acquisition. The latter is particularly pertinent where the acquisition of $K^+$ is concerned due to the physicochemical similarities between $Na^+$ and $K^+$ [15]. The capacity of plants to counteract salinity stress will strongly depend on the status of their $K^+$ nutrition. Conversely, although most plants can cope with external $K^+$ concentrations, however, physiological 'windows' of optimum $K^+$ concentrations narrows in the presence of increasing amounts of $Na^+$ [16].

*Calcium*
The application of potash as MOP or SOP increased the $Ca^{2+}$ uptake or availability of $Ca^{2+}$ at all selected experimental sites. However, the effect was more pronounced in the SOP treatments (Fig. 4). $Ca^{2+}$ is a major essential element and plays an important role in maintaining cell membrane stability and permeability. It increases germination and growth activities and speeds up the activities of enzymes involved in mitosis, cell division and elongation. It is also important for protein synthesis and carbohydrate transfer, and its presence may serve to detoxify the presence of heavy metals in plants. The availability of $Ca^{2+}$ is necessary for optimum growth and maintenance of metabolic activities in plants growing under stress conditions. Under saline conditions $Ca^{2+}$ compete with $Na^+$ for common uptake sites and many reports indicate that its uptake is hindered due to salinity [17]. However, the effect of SOP was found to be beneficial in the present study, in increasing either availability or uptake of $Ca^{2+}$. It is reported that higher Ca:Na ratios promote plant growth, which is very true in the present study as the plants at NIAB and Samundri had higher $Ca^{2+}$ and high growth parameters were recorded there. The $K^+$ played a major role in the enhancement of $Ca^{2+}$.

The amendment with organic manures significantly enhanced the $Ca^{2+}$ uptake. Maximum $Ca^{2+}$ concentrations were recorded in the plants treated with sugarcane press mud, closely followed by farmyard manure (FYM). But at NIAB, $Ca^{2+}$ was slightly higher in FYM than that of sugarcane press mud. It is a well established fact that organic manures are not only good sources of organic matter but they also contain some mineral nutrients [6]. So to have a good crop yield, amendment with organic manure is necessary.

*Phosphorous*
Potash application (SOP and MOP) enhanced the uptake of P; however, the effect of SOP was more pronounced than that of MOP (Fig. 4). The increase in P due to SOP may be due to the growth promoting effect of K and $SO_4$. Phosphorous is a major essential element necessary for optimum plant growth. It is also a component of certain enzymes and proteins, like adenosine triphosphate (ATP), ribonucleic acids (RNA), DNA and phytin. So its optimum uptake is necessary to maintain growth and metabolism. Its deficiency can affect the DNA and RNA synthesis, which effect not only growth but also genetic signals, required to pass on to seed, and causes

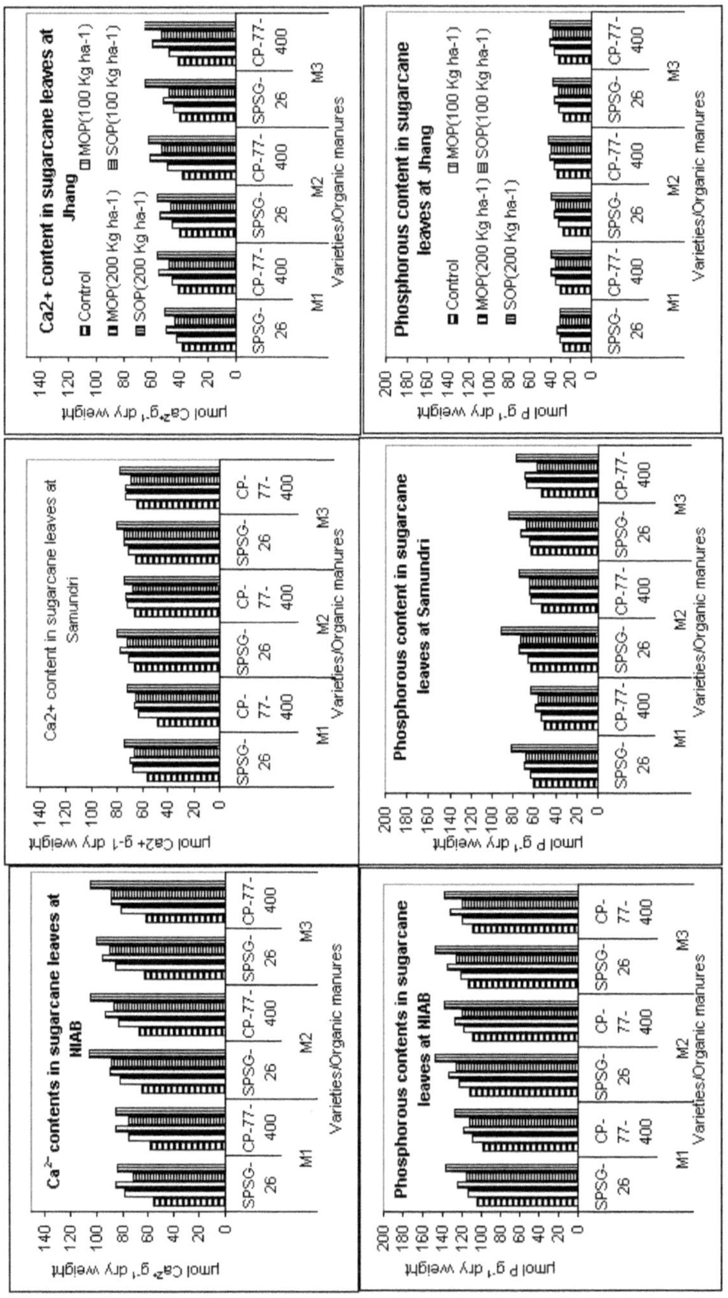


**Figure 4.** Interactive effect of SOP, MOP and organic manures on $Ca^{2+}$ and phosphorous contents in sugarcane leaves at different selected sites under saline and non-saline conditions

abnormal plant growth in the plants developed from P deficient seeds [18]. The present study revealed that plants with low P had low growth. However, treatment of SOP enhanced P uptake significantly and promoted growth and yield. Amendment with organic manure non-significantly enhanced uptake of P. Both the sources are beneficial in enhancing the P content in soil as well as in plants. From the results it can be concluded that the proper combination of SOP and organic manure can increase the yield of sugarcane and uptake of P.

*Nitrogen*

Nitrogen concentration reduced with salinity. Highest N contents were recorded in plants growing under normal conditions, followed by Samundri and Jhang (Fig. 5). Although salinity at Samundri was maximum at the time of sowing, the constant use of gypsum with water reduced it, and plant growth and nutrient uptake was better there than all other selected sites. It is well documented that salinity reduces the uptake of N; however, application of K enhances N in plants. Salinity may reduce the synthesis of certain enzymes involved in the nitrogen metabolism, such as nitrate reductase, or may reduce the substrate due to the presence of high salts in the root medium.

The data clearly indicated that the application of K significantly increased the uptake of nitrogen. Nitrogen uptake was enhanced by both sources, i.e., SOP and MOP; however, SOP was more efficient in increasing nitrogen concentrations than that of MOP. It is well known that nitrogen and potassium are plant nutrients required in high amounts for good crop growth and high production. Nitrogen is an essential element of biomolecules such as amino acids, proteins, nucleic acids, phytohormones and a number of enzymes and coenzymes. Nitrogen is deeply involved in the first step of growth which means the replication of chromosomes and which in biochemical terms consists mainly of the synthesis of deoxyribonucleic acids and nuclear protein [19]. On the other hand, potassium has no particular function as a constitutive element of biomolecules. Nevertheless, it is also an essential element for plants. Numerous physiological processes are known in which $K^+$ is involved and it was found that nitrogen and $K^+$ influence plant growth in a synergistic way. $K^+$ uptake into cells may contribute to the osmotic potential of the cytoplasm, which is a requisite of the osmotic water uptake. It is in the literature that $K^+$ has an activating effect on ATPase. It also contributes to depolarize the membrane by virtue of which ATPase activity increases [20, 21]. The permanent supply of $K^+$ to meristematic tissues stimulates the growth processes. This is why the plant treated with $K^+$ had better growth. In the growth process, phytohormones and $K^+$ are involved in a synergistic way. Meristematic growth is the prerequisite of crop production. It is for this reason that a balanced $K^+$ and N supply is of such a high relevance for having high yield.

The results about varietal performance showed that SPSG-26 took up N at NIAB and Samundri more efficiently than Jhang, where CP-77-400 showed better performance. Although SPSG-26 is salt tolerant (up to 10 $dSm^{-1}$), the plots selected for this variety had very high salinity and no additional gypsum treatment was imposed at these two places due to which SPSG-26 did not perform well.

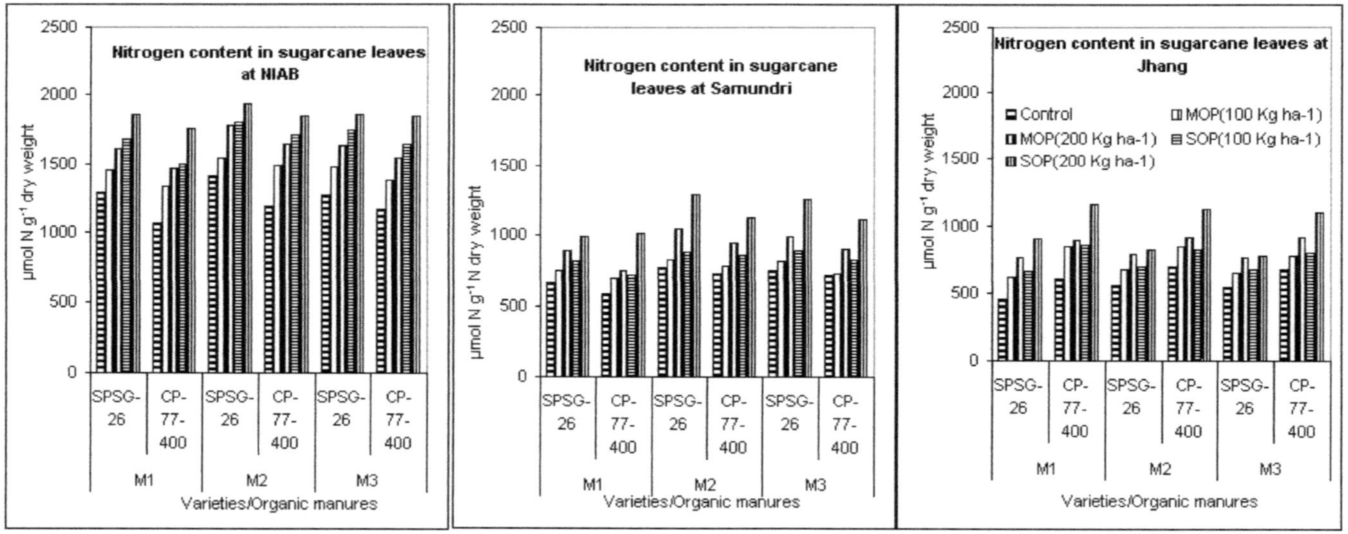


**Figure 5.** Interactive effect of SOP, MOP and organic manures on nitrogen contents in sugarcane leaves at different selected sites under saline and non-saline conditions

The amendment of organic manures also increased the uptake of nitrogen in both the varieties. The farmyard manure showed pronounced effects on nitrogen uptake. The organic manures are good sources of nitrogen and the importance of nitrogen in growth and other metabolic processes have been discussed above. From these results it can be concluded that for better plant growth, high yield and high nutrient availability to plants, a suitable combination of organic manure, nitrogen and potassium is required.

## Conclusion

From the results of the present study, it can be concluded that for optimum sugarcane growth and yield, balanced amount of $K^+$ and N in combination with organic manures are required. The fertilizer rates of this nutrient should be 50 % more for saline soil as compared to normal ones. For optimum sugarcane yield, the salinity of the soil selected for sugarcane cultivation should not be more than 10 dS $m^{-1}$, otherwise even tolerant varieties fail to produce economical yields. Selection of salt tolerant varieties is also of great importance in order to achieve higher sugarcane yields. SOP is more beneficial than MOP in many ways as has been discussed in this chapter.

## References


1 Anonymous (2004) Economic Survey 2003-04. Government of Pakistan, Finance Division, Economic Advisor Wing, Islamabad
2 Anonymous (2002) Economic Survey 2001-02. Government of Pakistan, Finance Division, Economic Advisor Wing, Islamabad
3 Kemmler G (1985) Soil fertility, fertilizer efficiency and crop production – A long-term view. Presented at $1^{st}$ National Congress of Soil Science, Lahore, Pakistan
4 Gorham I (1990) Salt tolerance in the triticeae. Ion discrimination in rye and *Triticales. J Expt Bot* 41: 609–614
5 Shah S, Gorham I, Forster BP, WynJones RG (1987) Salt tolerance in the triticeae: the contribution of the D genome to cation selectivity in hexaploid wheat. *J Expt Bot* 38: 254–269
6 Davis J, Daniel J (2002) Organic farming research project report submitted to the Organic Farming Research Foundation. Dept. of Soil and Crop Sciences Colorado State University, USA
7 Iqbal Z, Ashraf MY, Kashif M, Yamin M (2002) Use of industrial effluent ($CaCl_2$) for reclaiming salt-affected land and enhancing wheat yield. *Pakistan J Soil Sci* 21: 63–66
8 Nasir MN, Qureshi RH, Aslam M, Khtar J (2000) Screening of sugarcane lines selected through hydroponic studies in naturally salt-affected field. *Pak Sugar J* XV: 2–9
9 Dewis J, Freitas F (1970) Physical methods of soil and water analysis. *FAO. Soil Bull. No.10*, Rome, 39–51
10 Jackson ML (ed.) (1962) *Soil Chemical Analysis*. Constable and Company Ltd, UK
11 Watanabe FS, Olsen SR (1965) Test of an ascorbic acid method determining phosphorous in water extracts from soil. *Soil Sci Amer Proc* 29: 677–678
12 Ashraf MY, Wahed RA, Bhatti AS, Sarwar G, Aslam Z (1999) Salt tolerance potential in different Brassica species, Growth studies. In: H Hamdy, H Lieth, M Todorovic, M Moschenko (eds) In: *Halophytes Uses in Different Climates – II*. Backhuys Publishers, Leiden, The Netherlands, 119–125

13 Khan AH, Ashraf MY, Azmi AR (1992) Osmotic adjustment in sorghum under sodium chloride stress. *Act Physiol Plant* 14: 159–162
14 Munns R, Greenway HH, Kirst GO (1983) Halo-tolerant. In: OL Lang, PS Nobel, CB Osmond, H Ziegler (eds) In: *Physiological Plant Ecology III. Response to the Chemical and Biological Environment.* Encycl Plant Physiol New Ser, 12C: 59–135
15 Ashraf MY, Sarwar G (2002) Salt tolerance potential in some members of Brassicaceae. Physiological studies on water relations and mineral contents. In: R Ahmad, KA Malik (eds) In: *Prospects for Saline Agriculture.* Kluwer Academic Publishers, The Netherlands, 237–245
16 Ashraf MY, Khan AH, Azmi AR (1992) Cell membrane stability and its relation with some physiological process in wheat. *Acta Agron Hung* 41: 183–191
17 Khan AH, Ashraf MY, Naqvi SSM, Khanzada B, Ali M (1995) Growth, ion and solute contents of sorghum grown under NaCl and $Na_2SO_4$ salinity stress. *Acta Physiol Plant* 17: 261–268
18 Khan AH, Ashraf MY, Azmi AR (1993) Osmotic adjustment in wheat: A response to water stress. *Pak J Sci Ind Res* 36: 151–155
19 Ashraf M, Shabaz M, Ashraf MY (2001) Influence of nitrogen supply and water stress on growth and nitrogen, phosphorus, potassium and calcium content in pearlmillet. *Biologia Plantarum* 44: 459–462
20 Gorham I, McDonnell E, Budrewics E, Wyn Jones RG (1985) Salt tolerance in the Triticeae: Growth and solute accumulation in leaves of *Thinopyrum hessarabicum. J Expt Bot* 36: 1021–1030
21 Gorham I, Hardy C, WynJones RG, Joppa LR, Law CN (1987) Chromosomal location of a KINa discrimination character in the D genome of wheat. *Theoretical and Appl Gen* 74: 584–588

Biosaline Agriculture and Salinity Tolerance in Plants
Edited by M. Öztürk, Y. Waisel, M.A. Khan and G. Görk


# The effects of saline irrigation water by drip irrigation on salt distribution in soil

H. Hüsnü Kayikçioğlu and Bülent Okur

*Ege University Faculty of Agriculture, Department of Soil Science, Bornova–İzmir 35100, Turkey*

## Introduction

Depending on the increase in the salt amount in irrigation water in many arid and semi-arid regions of the world, salinity problems in agricultural soils appears and agricultural production is inhibited by excessive salt concentrations. This type of salinity occurs in topographical lowlands near the sea where intrusion of seawater to the aquifer is inevitable [1]. The reason why salinity is high in the areas near the sea, or in areas gained from the sea, is due to seawater intrusion. The salt composition of these soils is the same with that of the seawater [2]. The salinity of the seawater is accepted as approximately 0.35 %. The concentrations of main ions in seawater are 19.35 me $L^{-1}$ $Cl^-$, 10.752 me $L^{-1}$ $Na^+$, 2.701 me $L^{-1}$ $SO_4^=$, 1.295 me $L^{-1}$ $Mg^{+2}$ and 0.39 me $L^{-1}K^+$ [3]. Approximately 70 % of the world's population lives near, or at, the seaside. The rise in the population and the settlement in these areas increase the agricultural and industrial activities, hence a pressure on water sources takes place. The greatest effect of this pressure is on the change in the quality of underground water sources [4]. Various factors affect the salinity of underground water sources. The most important reason for the salinity in the aquifers at the seaside is the seawater intrusion [5].

Satsuma mandarin, which generally grow in the Gümüldür district of Izmir, Turkey, is economically important for this region. However in recent years, because of the seawater intrusion to the underground water sources, salinity problems in these soils have been of concern [6].

In the summer months, when mandarin orchards need irrigation, domestic water consumption increases also due to the rise in tourism. Further, seawater enters in the discharged underground water sources. For this reason, salinity stress in this region has serious effects on plant growth in the months of August and September.

The objective of the present study was to investigate the distribution of salinity using saline water with drip irrigation.

## Material and methods

Soil samples were taken from an experimental site which had been established in 1996 to determine the effects of salinity on yield and quality of Satsuma mandarins (cv. Owari) budded onto *Poncirus trifoliata* and *Troyer citrange* rootstocks at Ege University Campus in Izmir, Turkey. The tree spacing was 3 m between rows and 2.5 m on rows. Five different levels of irrigation with saline water (0.65 (fresh water-$I_0$), 2.00 ($I_1$); 3.50 ($I_2$); 5.00 ($I_3$) and 6.50 ($I_4$) dS $m^{-1}$) was realized during the summer months of 1996–2000 by drip irrigation [6]. Treatment plots were randomly located in each of four replicate blocks. Recommended amounts of nitrogen (N), phosphorus (P) and potassium (K) were applied. Soil samples were taken from the experimental field two times, on 16 August 2000 and 31 October 2000. Although the effective precipitation in the region up to the first sampling time did not occur, in October (which was the second sampling time), an average of 63 mm precipitation fell.

The drips with 2.3 $Lh^{-1}$ flow rate had the pressure regulated system and online type. Each different salt level contains different subjects ($I_0$, $I_1$, $I_2$, $I_3$ and $I_4$) and each subject contains three trees. There are a total of 240 mandarin trees in this experiment. The amounts of water and salt applied between the years 1996–2000 are given in Table 1, and the precipitation in Table 2. Soil samples were taken from two different depths (0–20 cm and 20–40 cm), and three different distances from the plant stem (0–20, 20–40 and 40–60 cm) in order to study clearly the salt distribution in the soil. The sampling was made for between (BR) and on rows (OR). Some physical and chemical properties of the experimental soil are given in Table 3 [6]. Soil samples were air dried and sieved (2 mm) before analyses [7]. Soil texture was determined by hydrometer method [8] and soil reaction and conductivity were measured in a 1:2.5 (W/W) aqueous solution [9]. Total $CaCO_3$ was assessed by using Scheibler calcimeter [10]. Furthermore, in the saturation extracts of soils pH [9], Na [9] and Cl [7] were determined. TARIS program was used in the statistic analysis of all data.

**Table 1.** The amounts of water and salt applied between 1996 and 2000

| Years | Beginning and Ending Date of Irrigation | Total Applied Water | | Total Applied Salt (kg) |
|---|---|---|---|---|
| | | (mm) | (liter) | |
| 1996 | 31 May – 15 November | 695.30 | 166,872 | – |
| 1997 | 27 May – 23 October | 517.36 | 124,166 | 163.50 |
| 1998 | 18 June – 12 October | 521.19 | 125,086 | 160.00 |
| 1999 | 11 June – 12 November | 443.52 | 106,445 | 158.00 |
| 2000 | 01 June – 14 November | 467.22 | 112,132 | 166.52 |

**Table 2.** Precipitation between 1996 and 2000 (mm)

| Years | 1996 | 1997 | 1998 | 1999 | 2000 |
|---|---|---|---|---|---|
| Total Year | 548.50 | 616.20 | 839.40 | 620.60 | 530.20 |
| Average Month | 45.71 | 51.35 | 69.95 | 51.72 | 44.18 |

**Table 3.** Some physical and chemical properties of the experiment soil in 1996

| pH ($H_2O$) | Total $CaCO_3$ (%) | $EC_e$ ($\mu S\ cm^{-1}$) | Sand (%) | Silt (%) | Clay (%) | Texture |
|---|---|---|---|---|---|---|
| 7.35 | 1.16 | 550 | 66.88 | 19.84 | 13.28 | Sandy loam |

## Results and discussion

Some chemical properties of the soils taken from 0–30 cm depth, and from 30 cm distance from the plant stem as control treatment in the first period are given in Table 4. The results that belong to the control parcel are accepted as the same for both *Poncirus trifoliata* and *Troyer citrange* rootstocks.

According to these results, the pH of the control parcel was analyzed as 5.84 and 4.57 in the direction of between and on rows, respectively. Total soluble salts in the water (%) did not change in relation to directions and was determined as 0.201 % on average. The electrical conductivities of the saturation extracts were found to be 3,550 $\mu S\ cm^{-1}$ and 3,650 $\mu S\ cm^{-1}$ between and on rows, respectively. The saturation extract of the control treatment contained an average of 2.95 me $l^{-1}$ $Na^+$ and 15.48 me $l^{-1}$ $Cl^-$.

Results related to the first period showed a linear increase in electrical conductivity of the saturation extract ($EC_e$) as well as the amount of soluble salts (%) at 0–20 cm soil depth depending upon the amount of added salt (Figs. 1a, 1b, 2a, 2b).

The amounts of soluble salts and $EC_e$ values on rows were higher than that of between rows because of the more salt accumulation on the laterals between drips. In addition, according to the results obtained from both rootstocks, the $EC_e$ values in both depths decreased as the distance from the plant stem increased (Tabs. 4 and 5; Figs. 1a, 1b).

Parallel to the increase in the salinity, the increases were also observed in the concentration of $Na^+$ and $Cl^-$ ions in the soil saturation extract. $Na^+$ concentrations of soils taken from 0–20 cm depth and from on rows of *Poncirus trifoliata* were determined as 16.96, 25.99, 29.98 and 31.75 me $l^{-1}$ depending on the enhanced salt rates (Figs. 3a and 3b). Similarly, $Cl^-$ concentrations were found to be 31.96, 34.94, 44.97 and 52.94 me $l^{-1}$ (Tab. 4; Figs. 4a and 4b).

In *Troyer citrange* rootstock these values were found to be 23.21, 24.93, 30.97 and 35.23 me $l^{-1}$ for $Na^+$ and 31.96, 36.96, 46.94 and 54.97 me $l^{-1}$ for $Cl^-$ (Tab. 5; Figs. 3a, 3b, 4a, 4b). In the saturation extract of the control soil, the $Na^+$ and $Cl^-$ concentration were 2.83 and 15.00 me $l^{-1}$ on rows and 3.06 and 15.96 me $l^{-1}$ between rows, respectively. The reason for the difference between $Na^+$ and $Cl^-$ concentrations of both rootstocks at the same treatment resulted from the different

**Table 4.** Chemical properties of *Poncirus trifoliata* soils (first sampling period)

| Treatment | Direction* | Depth (cm) | Distance (cm) | Total Soluble Salt (%) | pH (1:2.5) | Saturation extract | | | |
|---|---|---|---|---|---|---|---|---|---|
| | | | | | | $pH_e$ | $EC_e$ | me $l^{-1}$ | |
| | | | | | | | µS $cm^{-1}$ | $Na^+$ | $Cl^-$ |
| Control | OR | 30 | 0–30 | 0.200 | 5.84 | 6.47 | 3,550 | 2.83 | 15.00 |
| | BR | 30 | 0–30 | 0.202 | 4.57 | 6.59 | 3,650 | 3.06 | 15.96 |
| $I_1$(2.0 dS $m^{-1}$) | BR | 20 | 0–20 | 0.245 | 6.32 | 6.85 | 4,800 | 17.00 | 30.90 |
| | BR | 20 | 20–40 | 0.205 | 5.15 | 5.70 | 3,710 | 10.20 | 26.96 |
| | BR | 20 | 40–60 | 0.204 | 7.37 | 6.76 | 3,640 | 5.87 | 25.97 |
| | BR | 40 | 0– 20 | 0.212 | 7.16 | 6.69 | 3,365 | 11.31 | 26.98 |
| | BR | 40 | 20–40 | 0.195 | 7.35 | 6.63 | 3,075 | 10.44 | 22.97 |
| | BR | 40 | 40–60 | 0.200 | 7.36 | 6.68 | 3,442 | 8.48 | 20.86 |
| | OR | 20 | 0–20 | 0.249 | 6.99 | 6.83 | 5,010 | 16.96 | 31.96 |
| | OR | 20 | 20–40 | 0.220 | 7.37 | 6.71 | 4,260 | 13.27 | 28.96 |
| | OR | 20 | 40–60 | 0.188 | 7.46 | 6.83 | 2,880 | 6.31 | 20.97 |
| | OR | 40 | 0–20 | 0.200 | 7.08 | 6.81 | 2,910 | 8.87 | 20.98 |
| | OR | 40 | 20–40 | 0.203 | 7.29 | 6.88 | 3,702 | 11.31 | 25.97 |
| | OR | 40 | 40–60 | 0.165 | 7.45 | 6.84 | 2,910 | 6.52 | 19.97 |
| $I_2$(3.5 dS $m^{-1}$) | BR | 20 | 0–20 | 0.350 | 5.70 | 6.52 | 5,940 | 25.10 | 36.95 |
| | BR | 20 | 200–40 | 0.298 | 6.98 | 6.85 | 4,800 | 17.02 | 31.96 |
| | BR | 20 | 40–60 | 0.180 | 7.30 | 6.91 | 2,830 | 4.39 | 20.98 |
| | BR | 40 | 0–20 | 0.313 | 6.37 | 7.01 | 5,210 | 18.93 | 35.93 |
| | BR | 40 | 20–40 | 0.300 | 7.04 | 6.66 | 4,650 | 12.88 | 28.96 |
| | BR | 40 | 40–60 | 0.205 | 4.84 | 7.17 | 3,300 | 5.00 | 18.96 |
| | OR | 20 | 0–20 | 0.350 | 7.35 | 5.96 | 5,990 | 25.99 | 34.94 |
| | OR | 20 | 20–40 | 0.320 | 6.66 | 6.85 | 5,450 | 18.36 | 33.95 |
| | OR | 20 | 40–60 | 0.211 | 7.28 | 6.79 | 4,070 | 9.57 | 26.97 |
| | OR | 40 | 0–20 | 0.285 | 6.69 | 6.89 | 5,100 | 21.10 | 32.94 |
| | OR | 40 | 20–40 | 0.237 | 6.92 | 6.67 | 3,810 | 11.74 | 26.96 |
| | OR | 40 | 40–60 | 0.237 | 7.21 | 6.64 | 3,790 | 8.26 | 25.97 |
| $I_3$(5.0 dS $m^{-1}$) | BR | 20 | 0–20 | 0.380 | 4.85 | 6.20 | 6,950 | 28.26 | 44.93 |
| | BR | 20 | 20–40 | 0.310 | 6.85 | 6.88 | 5,180 | 20.23 | 34.95 |
| | BR | 20 | 40–60 | 0.202 | 7.30 | 7.23 | 3,445 | 11.53 | 22.97 |
| | BR | 40 | 0–20 | 0.330 | 6.13 | 6.91 | 5,350 | 25.66 | 36.95 |
| | BR | 40 | 20–40 | 0.245 | 7.11 | 7.40 | 4,400 | 16.22 | 33.97 |
| | BR | 40 | 40–60 | 0.210 | 7.42 | 6.92 | 3,382 | 8.70 | 21.96 |
| | OR | 20 | 0–20 | 0.385 | 4.81 | 6.19 | 7,190 | 29.98 | 44.97 |
| | OR | 20 | 20–40 | 0.320 | 5.23 | 6.29 | 5,620 | 25.23 | 37.96 |
| | OR | 20 | 40–60 | 0.238 | 6.94 | 6.76 | 4,180 | 14.35 | 28.96 |
| | OR | 40 | 0–20 | 0.345 | 6.14 | 6.56 | 5,880 | 23.27 | 39.96 |
| | OR | 40 | 20–40 | 0.290 | 6.84 | 6.98 | 4,850 | 19.31 | 28.96 |
| | OR | 40 | 40–60 | 0.215 | 7.40 | 7.03 | 3,560 | 12.40 | 21.97 |
| $I_4$(6.5 dS $m^{-1}$) | BR | 20 | 0–20 | 0.400 | 6.77 | 6.82 | 7,400 | 31.75 | 51.93 |
| | BR | 20 | 20–40 | 0.310 | 7.34 | 6.78 | 5,580 | 25.75 | 41.96 |
| | BR | 20 | 40–60 | 0.220 | 7.37 | 7.02 | 4,010 | 10.74 | 30.95 |
| | BR | 40 | 0–20 | 0.318 | 7.35 | 6.97 | 5,450 | 23.79 | 40.94 |
| | BR | 40 | 20–40 | 0.260 | 7.47 | 6.98 | 4,120 | 16.09 | 30.95 |
| | BR | 40 | 40–60 | 0.225 | 7.43 | 6.97 | 3,541 | 7.61 | 27.00 |
| | OR | 20 | 0–20 | 0.410 | 6.51 | 7.37 | 7,520 | 31.75 | 52.94 |
| | OR | 20 | 20–40 | 0.362 | 7.02 | 7.06 | 6,110 | 27.40 | 42.94 |
| | OR | 20 | 40–60 | 0.242 | 7.45 | 7.32 | 3,840 | 9.53 | 31.96 |
| | OR | 40 | 0–20 | 0.365 | 7.23 | 7.16 | 5,940 | 25.23 | 39.97 |
| | OR | 40 | 20–40 | 0.308 | 7.30 | 7.49 | 5,065 | 21.31 | 35.95 |
| | OR | 40 | 40–60 | 0.220 | 7.50 | 7.5 | 3,440 | 8.15 | 25.83 |

* ***BR:*** *Between rows direction* * ***OR:*** *On rows direction*

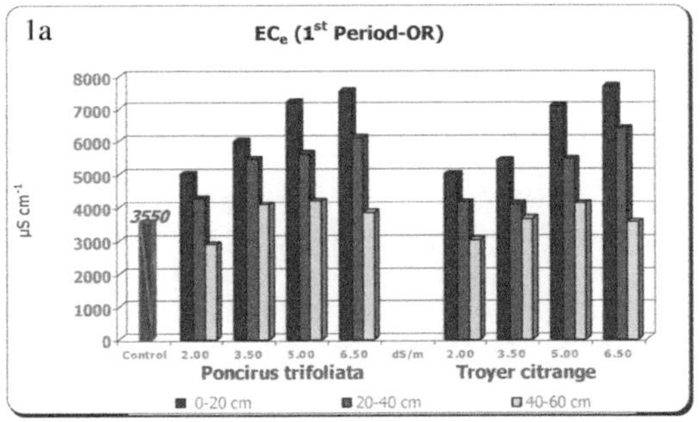


**Figure 1a.** $EC_e$ values at 0–20 cm depth of soils taken on rows (OR) in the first period

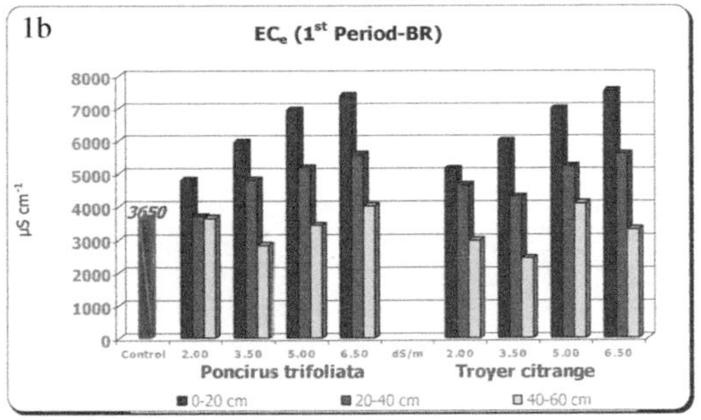


**Figure 1b.** $EC_e$ values at 0–20 cm depth of soils taken between rows (BR) in the first period

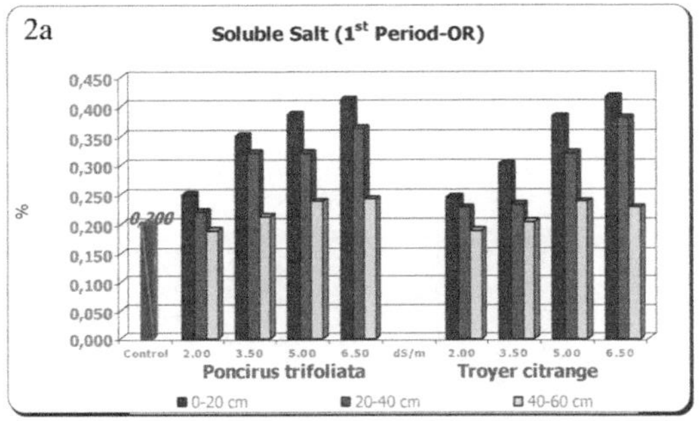


**Figure 2a.** Amount of total soluble salt at 0–20 cm depth of soils taken on rows (OR) in the first period

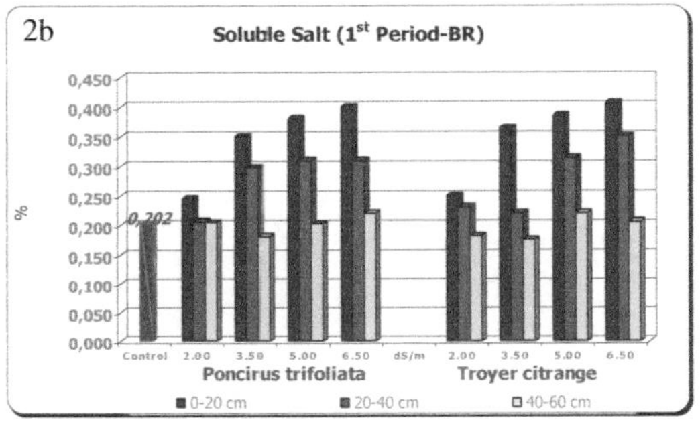


**Figure 2b.** Amount of total soluble salt at 0–20 cm depth of soils taken between rows (BR) in the first period

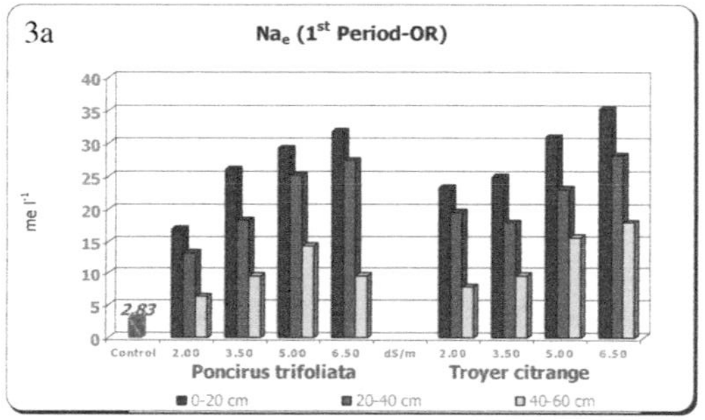


**Figure 3a.** $Na_e^+$ values at 0–20 cm depth of soils taken on rows (OR) in the first period

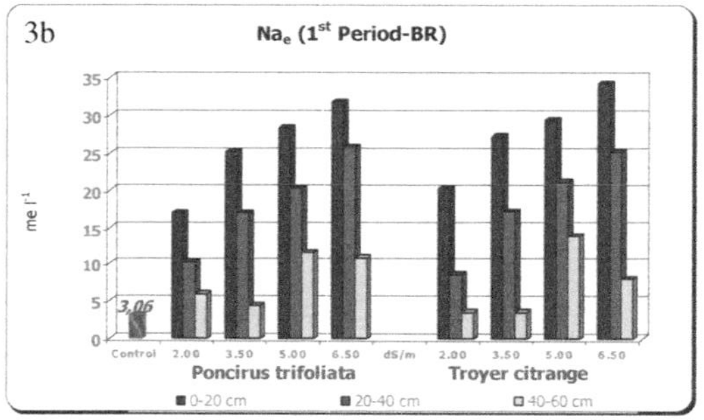


**Figure 3b.** $Na_e^+$ values at 0–20 cm depth of soils taken between rows (BR) in the first period

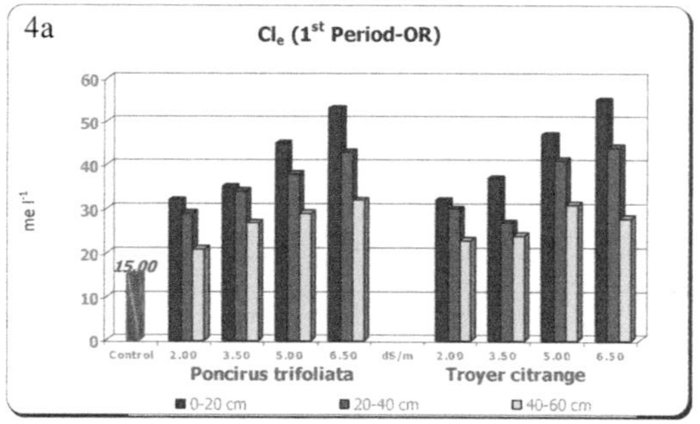


**Figure 4a.** $Cl_e^-$ values at 0–20 cm depth of soils taken on rows (OR) in the first period

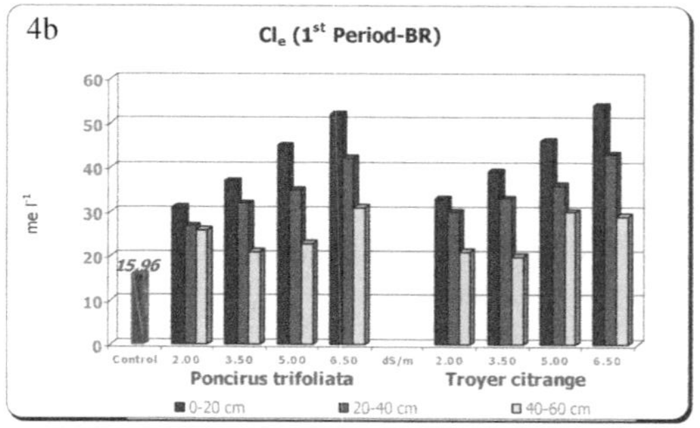


**Figure 4b.** $Cl_e^-$ values at 0–20 cm depth of soils taken between rows (BR) in the first period

characteristic of the rootstocks [6, 11]. The increase in the $Na_e$ and $Cl_e$ in relation to increased salinity at the first depth and distance also occurred at the second distance. But this significant increase could not be determined at the farthest distance (40–60 cm) from the plant stem in each salt treatment (Figs 3a, 3b, 4a, 4b).

In the soils of the second period, salinity increased as the distance of drippers and the soil depth increased. Nevertheless, in these soils where more saline irrigation water was applied, the amount of total soluble salt, $EC_e$, $Na_e^+$ and $Cl_e^-$ decreased in the soil under the soil surface (Tabs. 6 and 7). The increase of these parameters depending on the increased soil depth showed that the leaching occurred under the surface. It can be said that an average of 63 mm of rain fell during the second sampling period, and from October leached the salts to deeper zones and carried throughout the laterals.

$Na_e$ and $EC_e$ values at 0–20 cm soil depth, just under the drippers significantly decreased compared to the first period. However, these parameters clearly increased at the distances of 0–20 cm and 20–40 cm from the stem in both directions (Figs. 5a, 5b, 7a, 7b). The characteristic leaching area occurred just under the drippers caused by the moving of salts from this region and accumulating in the outside of the leaching area throughout laterals (Figs. 6a and 6b).

In the second period soils, the concentration of $Cl^-$ ion increased in all directions (OR and BR) and distances (0–20, 20–40 and 40–60 cm) for all salt treatments except for $I_1$ treatment between rows at the first distance compared to the first period (Figs. 8a and 8b).

The fact that $Na^+$ and $Cl^-$ could not leach deeper soil layers and ions move to upper layers by capillarity in arid conditions, the concentrations of $Na^+$ and $Cl^-$ decreased at 20–40 cm soil depth of the first period. Whereas in the soils sampled in the second period, higher amounts of saline water were applied, the leaching effect of rain occurred and water movement became slow by capillarity. For this reason, the leaching of salts from the soil surface occurred in higher levels than the first period soils and the concentrations of $Na^+$ and $Cl^-$ at the second depth clearly increased in both rootstocks compared to the first period.

Irrigating the soils with more saline water in the second period, less salt under the drips which is a characteristic of drip irrigation systems and 63 mm rainfall in October 2000 affected the distribution of $Cl^-$and $Na^+$ ions, as considered above.

Similarly to $Na^+$ ion, $Cl^-$ ion also increased between and on rows of two of the rootstocks.

The decrease of soil reaction in the second period could have resulted from $HNO_3$ used for adjusting the pH of nutrient solution and for preventing plugging of laterals and drippers used in the irrigation system. While the pH of the first period soils varied from 4.50–7.52, in the second period these values were in the range of 5.12–7.22.

In the first period samples, while salt accumulation occurred at 0–20 cm soil depth and 0–20 cm distance from the stem between and on rows, in the second period samples, salt moved to 20–40 and 40–60 cm distances from the stem and to 20–40 cm soil depth. For this reason, salt accumulation occurred at a greater distance from the drippers. In fact, salt accumulates outside the wetting area in drip irrigation.

**Table 5.** Chemical properties of *Troyer citrange* soils (first sampling period)

| Treatment | Direction* | Depth (cm) | Distance (cm) | Total Soluble Salt (%) | pH (1:2.5) | Saturation extract | | | |
|---|---|---|---|---|---|---|---|---|---|
| | | | | | | $pH_e$ | $EC_e$ $\mu S\ cm^{-1}$ | $me\ l^{-1}$ $Na^+$ | $Cl^-$ |
| $I_1$ (2.0 dS $m^{-1}$) | BR | 20 | 0–20 | 0.250 | 4.50 | 6.36 | 5,150 | 20.24 | 32.95 |
| | BR | 20 | 20–40 | 0.230 | 6.78 | 7.24 | 4,654 | 8.51 | 29.97 |
| | BR | 20 | 40–60 | 0.181 | 7.18 | 7.51 | 2,980 | 3.48 | 20.98 |
| | BR | 40 | 0–20 | 0.210 | 6.71 | 7.40 | 3,950 | 12.96 | 27.95 |
| | BR | 40 | 20–40 | 0.188 | 7.15 | 7.35 | 3,150 | 7.16 | 23.83 |
| | BR | 40 | 40–60 | 0.175 | 7.10 | 7.27 | 3,060 | 6.06 | 19.98 |
| | OR | 20 | 0–20 | 0.245 | 4.55 | 5.81 | 5,000 | 23.21 | 31.96 |
| | OR | 20 | 20–40 | 0.227 | 5.74 | 7.14 | 4,150 | 19.54 | 29.96 |
| | OR | 20 | 40–60 | 0.188 | 7.34 | 7.57 | 3,010 | 7.74 | 22.97 |
| | OR | 40 | 0–20 | 0.232 | 6.71 | 7.45 | 3,750 | 10.92 | 27.95 |
| | OR | 40 | 20–40 | 0.230 | 7.10 | 7.40 | 3,640 | 8.64 | 26.95 |
| | OR | 40 | 40–60 | 0.203 | 7.39 | 7.33 | 3,260 | 7.16 | 20.97 |
| $I_2$ (3.5 dS $m^{-1}$) | BR | 20 | 0–20 | 0.365 | 5.14 | 6.25 | 6,010 | 27.12 | 38.96 |
| | BR | 20 | 20–40 | 0.220 | 6.81 | 7.40 | 4,300 | 17.21 | 32.97 |
| | BR | 20 | 40–60 | 0.175 | 7.36 | 7.59 | 2,450 | 3.48 | 19.96 |
| | BR | 40 | 0–20 | 0.243 | 6.66 | 7.30 | 4,760 | 17.99 | 33.86 |
| | BR | 40 | 20–40 | 0.232 | 7.00 | 7.27 | 4,550 | 19.54 | 25.83 |
| | BR | 40 | 40–60 | 0.162 | 7.43 | 7.58 | 2,645 | 8.51 | 20.98 |
| | OR | 20 | 0–20 | 0.303 | 4.99 | 6.51 | 5,420 | 24.93 | 36.96 |
| | OR | 20 | 20–40 | 0.232 | 6.03 | 6.80 | 4,090 | 17.96 | 26.97 |
| | OR | 20 | 40–60 | 0.203 | 7.15 | 6.67 | 3,650 | 9.53 | 23.98 |
| | OR | 40 | 0–20 | 0.260 | 6.39 | 6.79 | 4,930 | 21.29 | 29.89 |
| | OR | 40 | 20–40 | 0.203 | 6.87 | 7.00 | 3,340 | 17.23 | 23.96 |
| | OR | 40 | 40–60 | 0.170 | 7.40 | 7.17 | 2,640 | 9.18 | 20.98 |
| $I_3$ (5.0 dS $m^{-1}$) | BR | 20 | 0–20 | 0.385 | 5.39 | 6.55 | 7,010 | 29.36 | 45.94 |
| | BR | 20 | 20–40 | 0.314 | 6.69 | 7.52 | 5,220 | 21.18 | 35.87 |
| | BR | 20 | 40–60 | 0.220 | 7.34 | 7.66 | 4,100 | 13.80 | 29.98 |
| | BR | 40 | 0–20 | 0.358 | 6.65 | 7.26 | 6,010 | 27.40 | 40.96 |
| | BR | 40 | 20–40 | 0.290 | 6.98 | 7.23 | 5,120 | 20.66 | 32.97 |
| | BR | 40 | 40–60 | 0.220 | 7.19 | 7.23 | 4,050 | 11.70 | 26.96 |
| | OR | 20 | 0–20 | 0.382 | 5.18 | 6.35 | 7,075 | 30.97 | 46.94 |
| | OR | 20 | 20–40 | 0.320 | 7.04 | 7.27 | 5,450 | 23.05 | 40.97 |
| | OR | 20 | 40–60 | 0.238 | 7.29 | 7.32 | 4,110 | 15.66 | 30.97 |
| | OR | 40 | 0–20 | 0.350 | 6.13 | 7.11 | 5,910 | 29.58 | 42.94 |
| | OR | 40 | 20–40 | 0.280 | 6.97 | 7.32 | 4,655 | 22.47 | 32.94 |
| | OR | 40 | 40–60 | 0.218 | 7.17 | 7.30 | 4,000 | 14.15 | 27.97 |
| $I_4$ (6.5 dS $m^{-1}$) | BR | 20 | 20–40 | 0.350 | 7.08 | 6.94 | 5,590 | 25.10 | 42.94 |
| | BR | 20 | 40–60 | 0.205 | 7.48 | 7.09 | 3,310 | 7.96 | 28.97 |
| | BR | 40 | 0–20 | 0.374 | 6.83 | 7.19 | 6,180 | 32.59 | 42.96 |
| | BR | 40 | 20–40 | 0.354 | 7.20 | 7.10 | 5,640 | 29.80 | 36.95 |
| | BR | 40 | 40–60 | 0.272 | 7.42 | 7.21 | 4,120 | 20.16 | 27.94 |
| | OR | 20 | 0–20 | 0.415 | 7.10 | 6.71 | 7,670 | 35.23 | 54.97 |
| | OR | 20 | 20–40 | 0.380 | 6.90 | 7.06 | 6,380 | 28.20 | 43.97 |
| | OR | 20 | 40–60 | 0.227 | 7.52 | 7.19 | 3,550 | 17.96 | 27.96 |
| | OR | 40 | 0–20 | 0.377 | 6.53 | 7.02 | 6,170 | 27.09 | 43.94 |
| | OR | 40 | 20–40 | 0.342 | 7.06 | 7.00 | 5,642 | 24.65 | 37.95 |
| | OR | 40 | 40–60 | 0.280 | 7.48 | 7.16 | 4,370 | 18.56 | 31.94 |

* ***BR:** Between rows direction* * ***OR:** On rows direction*

**Table 6.** Chemical properties of *Poncirus trifoliata* soils (second sampling period)

| Treatment | Direction* | Depth (cm) | Distance (cm) | Total Soluble Salt (%) | pH (1:2.5) | Saturation extract | | | |
|---|---|---|---|---|---|---|---|---|---|
| | | | | | | $pH_e$ | $EC_e$ $\mu S\ cm^{-1}$ | $me\ l^{-1}$ $Na^+$ | $Cl^-$ |
| Control | OR | 30 | 0–30 | 0.200 | 6.97 | 7.27 | 3,250 | 4.40 | 15.83 |
| | BR | 30 | 0–30 | 0.205 | 6.02 | 6.88 | 3,400 | 3.67 | 14.96 |
| $I_1$ (2.0 $dS\ m^{-1}$) | BR | 20 | 0–220 | 0.281 | 6.67 | 7.55 | 4,320 | 12.73 | 26.97 |
| | BR | 20 | 20–40 | 0.320 | 7.01 | 7.40 | 5,010 | 13.97 | 29.97 |
| | BR | 20 | 40–60 | 0.334 | 7.15 | 7.49 | 5,210 | 18.51 | 31.97 |
| | BR | 40 | 0–20 | 0.180 | 7.08 | 7.68 | 3,105 | 12.32 | 21.97 |
| | BR | 40 | 20–40 | 0.281 | 7.04 | 7.69 | 4,440 | 13.82 | 24.97 |
| | BR | 40 | 40–60 | 0.294 | 6.71 | 7.59 | 4,864 | 15.97 | 28.97 |
| | OR | 20 | 0–20 | 0.276 | 6.17 | 7.39 | 4,150 | 13.97 | 28.96 |
| | OR | 20 | 20–40 | 0.348 | 7.15 | 7.54 | 5,123 | 15.62 | 29.96 |
| | OR | 20 | 40–60 | 0.348 | 7.21 | 7.48 | 5,390 | 16.27 | 31.97 |
| | OR | 40 | 0–20 | 0.180 | 6.57 | 7.65 | 3,056 | 11.67 | 22.96 |
| | OR | 40 | 20–40 | 0.262 | 7.06 | 7.63 | 4,290 | 11.02 | 23.97 |
| | OR | 40 | 40–60 | 0.330 | 7.22 | 7.56 | 5,390 | 18.60 | 29.96 |
| $I_2$ (3.5 $dS\ m^{-1}$) | BR | 20 | 20–40 | 0.392 | 6.59 | 7.03 | 6,840 | 24.78 | 44.96 |
| | BR | 20 | 40–60 | 0.349 | 6.73 | 7.12 | 5,520 | 22.41 | 40.91 |
| | BR | 40 | 0–20 | 0.335 | 6.52 | 7.01 | 5,285 | 21.49 | 35.92 |
| | BR | 40 | 20–40 | 0.389 | 6.60 | 6.82 | 6,245 | 23.04 | 42.92 |
| | BR | 40 | 40–60 | 0.401 | 6.39 | 7.09 | 6,854 | 25.18 | 45.93 |
| | OR | 20 | 0–20 | 0.300 | 6.53 | 6.84 | 4,670 | 20.19 | 34.91 |
| | OR | 20 | 20–40 | 0.349 | 6.46 | 6.96 | 5,640 | 21.63 | 37.96 |
| | OR | 20 | 40–60 | 0.383 | 6.82 | 7.23 | 6,240 | 23.20 | 44.96 |
| | OR | 40 | 0–20 | 0.307 | 6.18 | 7.16 | 4,820 | 21.20 | 35.92 |
| | OR | 40 | 20–40 | 0.392 | 5.79 | 7.30 | 6,100 | 24.10 | 42.87 |
| | OR | 40 | 40–60 | 0.360 | 6.68 | 6.96 | 5,780 | 22.20 | 44.83 |
| $I_3$ (5.0 $dS\ m^{-1}$) | BR | 20 | 0–20 | 0.381 | 5.12 | 6.36 | 6,340 | 25.79 | 46.91 |
| | BR | 20 | 20–40 | 0.381 | 5.54 | 6.72 | 6,800 | 26.15 | 49.92 |
| | BR | 20 | 40–60 | 0.368 | 5.76 | 6.60 | 6,350 | 23.20 | 43.96 |
| | BR | 40 | 0–20 | 0.393 | 5.18 | 6.12 | 7,010 | 28.10 | 51.97 |
| | BR | 40 | 20–40 | 0.393 | 5.63 | 6.77 | 7,050 | 28.56 | 52.93 |
| | BR | 40 | 40–60 | 0.406 | 5.87 | 6.98 | 7,200 | 30.04 | 55.92 |
| | OR | 20 | 0–20 | 0.361 | 6.05 | 7.29 | 6,250 | 23.18 | 44.97 |
| | OR | 20 | 20–40 | 0.357 | 5.94 | 7.22 | 6,105 | 26.42 | 48.96 |
| | OR | 20 | 40–60 | 0.387 | 5.92 | 7.00 | 6,802 | 29.18 | 56.97 |
| | OR | 40 | 0–20 | 0.372 | 5.65 | 6.90 | 6,640 | 26.20 | 48.92 |
| | OR | 40 | 20–40 | 0.400 | 6.29 | 6.75 | 7,215 | 30.64 | 56.91 |
| | OR | 40 | 40–60 | 0.387 | 6.46 | 6.82 | 6,980 | 29.00 | 55.91 |
| $I_4$ (6.5 $dS\ m^{-1}$) | BR | 20 | 0–20 | 0.440 | 7.14 | 7.45 | 7,120 | 29.18 | 55.83 |
| | BR | 20 | 20–40 | 0.465 | 7.06 | 7.36 | 7,460 | 30.41 | 59.91 |
| | BR | 20 | 40–60 | 0.465 | 7.07 | 7.53 | 7,500 | 32.13 | 63.87 |
| | BR | 40 | 0–20 | 0.431 | 7.16 | 7.50 | 7,010 | 27.28 | 52.94 |
| | BR | 40 | 20–40 | 0.473 | 7.10 | 7.50 | 7,650 | 33.51 | 64.92 |
| | BR | 40 | 40–60 | 0.479 | 7.14 | 7.62 | 7,715 | 37.20 | 68.81 |
| | OR | 20 | 0–20 | 0.420 | 6.89 | 7.39 | 7,010 | 27.25 | 52.87 |
| | OR | 20 | 20–40 | 0.423 | 6.95 | 7.45 | 7,195 | 29.64 | 57.91 |
| | OR | 20 | 40–60 | 0.430 | 6.92 | 7.60 | 7,220 | 28.80 | 55.93 |
| | OR | 40 | 0–20 | 0.429 | 7.00 | 7.56 | 7,073 | 27.43 | 53.81 |
| | OR | 40 | 20–40 | 0.460 | 6.92 | 7.68 | 7,495 | 34.01 | 62.93 |

** **BR:** Between rows direction * **OR:** On rows direction*

**Table 7.** Chemical properties of *Troyer citrange* soils (second sampling period)

| Treatment | Direction* | Depth (cm) | Distance (cm) | Total Soluble Salt (%) | pH (1:2.5) | Saturation extract | | | |
|---|---|---|---|---|---|---|---|---|---|
| | | | | | | $pH_e$ | $EC_e$ $\mu S\,cm^{-1}$ | me $l^{-1}$ $Na^+$ | $Cl^-$ |
| $I_1$ (2.0 dS $m^{-1}$) | BR | 20 | 0–20 | 0.260 | 6.79 | 7.65 | 4,500 | 18.98 | 28.93 |
| | BR | 20 | 20–40 | 0.275 | 6.83 | 7.68 | 4,430 | 15.20 | 28.94 |
| | BR | 20 | 40–60 | 0.293 | 6.90 | 7.70 | 4,850 | 19.30 | 32.94 |
| | BR | 40 | 0–20 | 0.190 | 6.80 | 7.52 | 3,450 | 15.73 | 26.96 |
| | BR | 40 | 20–40 | 0.263 | 6.73 | 7.50 | 4,220 | 16.70 | 27.94 |
| | BR | 40 | 40–60 | 0.297 | 6.92 | 7.65 | 4,860 | 18.10 | 30.97 |
| | OR | 20 | 0–20 | 0.210 | 6.29 | 7.29 | 3,790 | 16.53 | 25.91 |
| | OR | 20 | 20–40 | 0.290 | 6.46 | 7.35 | 4,680 | 18.25 | 29.83 |
| | OR | 20 | 40–60 | 0.290 | 6.37 | 7.38 | 4,795 | 19.10 | 30.96 |
| | OR | 40 | 0–20 | 0.200 | 6.37 | 7.30 | 3,550 | 16.23 | 25.93 |
| | OR | 40 | 20–40 | 0.270 | 6.52 | 7.62 | 4,320 | 17.43 | 29.94 |
| | OR | 40 | 40–60 | 0.300 | 6.40 | 7.52 | 4,886 | 18.24 | 32.96 |
| $I_2$ (3.5 dS $m^{-1}$) | BR | 20 | 0–20 | 0.365 | 6.52 | 7.49 | 6,053 | 19.20 | 39.83 |
| | BR | 20 | 20–40 | 0.360 | 6.41 | 7.42 | 5,920 | 20.49 | 40.83 |
| | BR | 20 | 40–60 | 0.323 | 6.37 | 7.63 | 5,460 | 21.86 | 41.89 |
| | BR | 40 | 0–20 | 0.371 | 6.29 | 7.56 | 6,110 | 21.18 | 42.97 |
| | BR | 40 | 20–40 | 0.374 | 6.46 | 7.60 | 6,250 | 22.10 | 44.96 |
| | BR | 40 | 40–60 | 0.390 | 6.39 | 7.60 | 6,362 | 23.20 | 45.87 |
| | OR | 20 | 0–20 | 0.303 | 6.54 | 7.40 | 4,750 | 20.38 | 34.96 |
| | OR | 20 | 20–40 | 0.335 | 6.29 | 7.53 | 5,696 | 24.03 | 36.93 |
| | OR | 20 | 40–60 | 0.370 | 6.70 | 7.46 | 6,256 | 25.20 | 42.97 |
| | OR | 40 | 0–20 | 0.300 | 6.49 | 7.39 | 5,069 | 24.32 | 38.96 |
| | OR | 40 | 20–40 | 0.372 | 6.84 | 7.56 | 6,272 | 27.20 | 44.97 |
| | OR | 40 | 40–60 | 0.349 | 6.60 | 7.40 | 5,897 | 26.18 | 41.96 |
| $I_3$ (5.0 dS $m^{-1}$) | BR | 20 | 0–20 | 0.374 | 5.90 | 7.26 | 6,725 | 26.96 | 48.91 |
| | BR | 20 | 20–40 | 0.379 | 6.05 | 7.00 | 6,805 | 28.19 | 51.92 |
| | BR | 20 | 40–60 | 0.381 | 6.20 | 7.10 | 6,890 | 24.49 | 50.92 |
| | BR | 40 | 0–20 | 0.385 | 6.00 | 7.22 | 6,985 | 29.38 | 53.92 |
| | BR | 40 | 20–40 | 0.390 | 6.10 | 7.26 | 7,010 | 28.90 | 56.91 |
| | BR | 40 | 40–60 | 0.398 | 6.19 | 7.29 | 7,150 | 31.18 | 57.91 |
| | OR | 20 | 0–20 | 0.352 | 6.25 | 7.30 | 6,500 | 25.94 | 46.92 |
| | OR | 20 | 20–40 | 0.370 | 6.30 | 7.51 | 6,653 | 26.95 | 51.92 |
| | OR | 20 | 40–60 | 0.390 | 6.25 | 7.46 | 6,948 | 29.38 | 54.92 |
| | OR | 40 | 0–20 | 0.370 | 6.32 | 7.53 | 6,648 | 27.40 | 49.92 |
| | OR | 40 | 20–40 | 0.405 | 6.29 | 7.51 | 7,250 | 29.90 | 60.92 |
| | OR | 40 | 40–60 | 0.400 | 6.34 | 7.60 | 7,175 | 30.05 | 57.92 |
| $I_4$ (6.5 dS $m^{-1}$) | BR | 20 | 0–20 | 0.420 | 6.35 | 7.58 | 7,000 | 29.96 | 53.96 |
| | BR | 20 | 20–40 | 0.440 | 6.56 | 7.45 | 7,310 | 30.86 | 57.96 |
| | BR | 20 | 40–60 | 0.469 | 6.27 | 7.60 | 7,553 | 34.91 | 68.96 |
| | BR | 40 | 0–20 | 0.435 | 6.35 | 7.34 | 7,115 | 29.44 | 54.93 |
| | BR | 40 | 20–40 | 0.490 | 6.46 | 7.39 | 7,785 | 36.19 | 68.83 |
| | BR | 40 | 40–60 | 0.450 | 6.50 | 7.50 | 7,480 | 37.30 | 66.87 |
| | OR | 20 | 0–20 | 0.400 | 6.43 | 7.28 | 7,015 | 27.86 | 54.97 |
| | OR | 20 | 20–40 | 0.427 | 6.60 | 7.37 | 7,151 | 29.44 | 58.81 |
| | OR | 20 | 40–60 | 0.443 | 6.29 | 7.20 | 7,511 | 36.20 | 61.86 |
| | OR | 40 | 0–20 | 0.417 | 6.37 | 7.35 | 7,108 | 28.88 | 55.87 |
| | OR | 40 | 20–40 | 0.468 | 6.50 | 7.45 | 7,550 | 37.64 | 64.96 |
| | OR | 40 | 40–60 | 0.441 | 6.50 | 7.00 | 7,384 | 32.47 | 62.96 |

** **BR:** Between rows direction * **OR:** On rows direction*

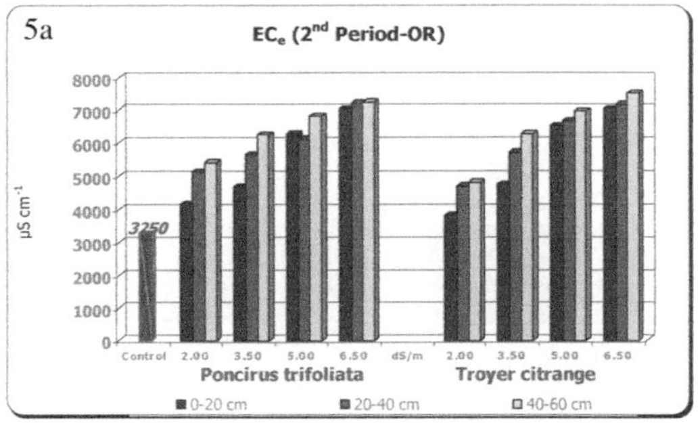


**Figure 5a.** $EC_e$ values at 0–20 cm depth of soils taken on rows (OR) in the second period

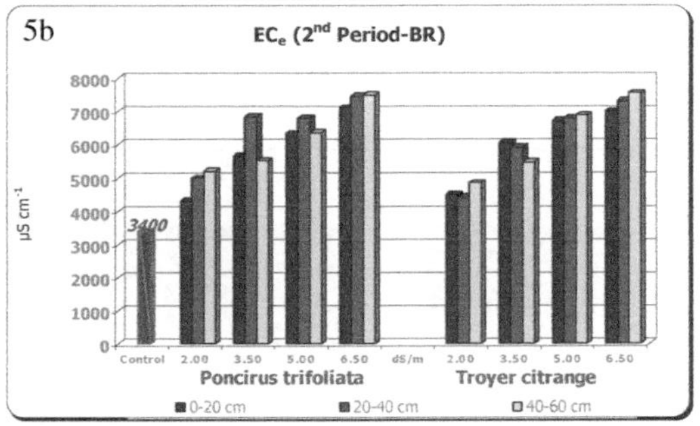


**Figure 5b.** $EC_e$ values at 0–20 cm depth of soils taken between rows (BR) in the second period

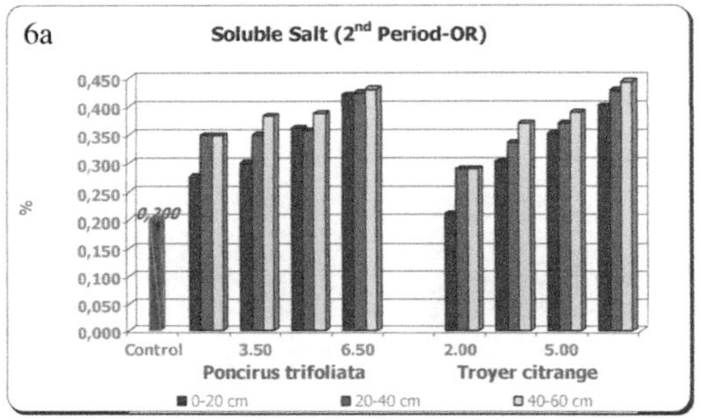


**Figure 6a.** Amount of total soluble salt at 0–20 cm depth of soils taken on rows (OR) in the second period

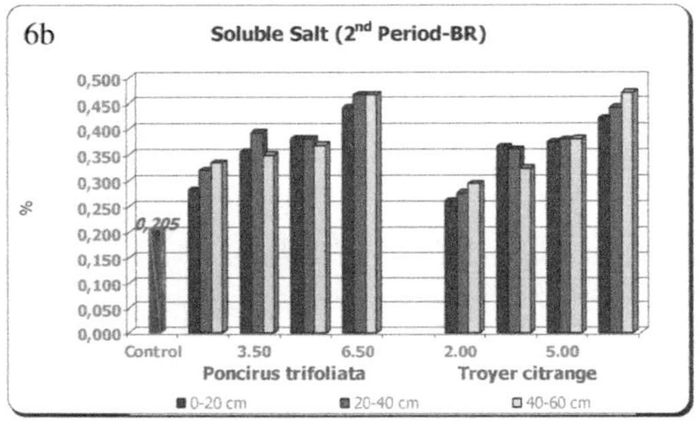


**Figure 6b.** Amount of total soluble salt at 0–20 cm depth of soils taken between rows (BR) in the second period

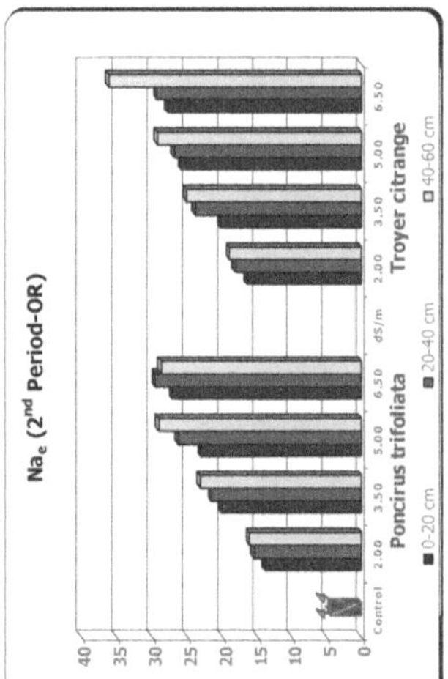


**Figure 7a.** $Na_e^+$ values at 0–20 cm depth of soils taken on rows (OR) in the second period

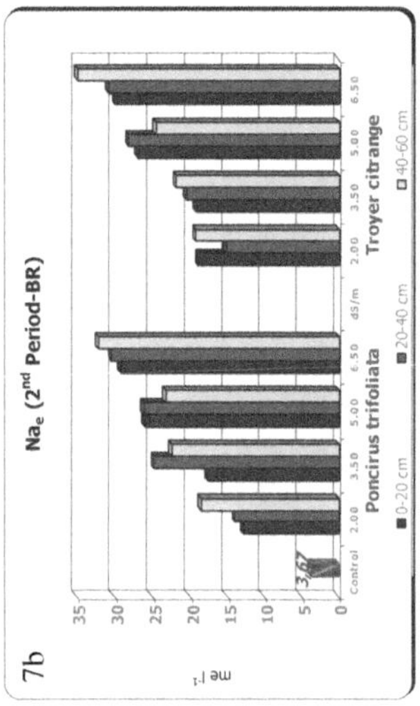


**Figure 7b.** $Na_e^+$ values at 0–20 cm depth of soils taken between rows (BR) in the second period

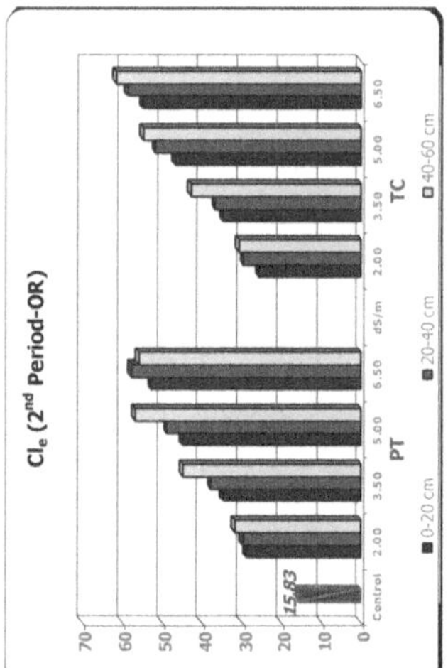


**Figure 8a.** $Cl_e^-$ values at 0–20 cm depth of soils taken on rows (OR) in the second period

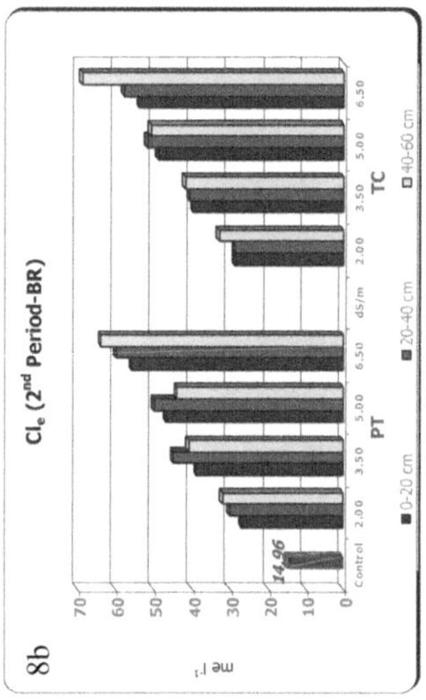


**Figure 8b.** $Cl_e^-$ values at 0–20 cm depth of soils taken between rows (BR) in the second period

The reason for the deeper leaching of salts was probably due to the 63 mm rainfall during October.

According to the variance analysis and LSD test, a 1 % significant relationship was found between the applied salt and total soluble salts, electrical conductivity of saturation extract ($EC_e$), $Na^+$ and $Cl^-$ in this extract.

The fact that the composition of salt applied to soils is the same with that of seawater (dominant salt NaCl), only the concentration of $Na^+$ and $Cl^-$ showed a linear increase depending on the increased salt levels.

Soil reaction depending on the salt levels was negative at 1 % significance level in the first period of soils of *Poncirus trifoliata*. The same change was positive at 1 % significance level in *Troyer citrange*. According to these results, it can be suggested that *Poncirus trifoliata* rootstocks, which can uptake more $Na^+$ ions from soils compared to *Troyer citrange* rootstock, can prevent the increase of pH due to $Na^+$. Also, the reason for the positive relationship between pH and the levels of applied salt in *Troyer citrange* can be due to this rootstocks significant uptake of $Na^+$ from the soil.

$Na^+$ concentrations in soils belonging to *Poncirus trifoliata* rootstock was less in both periods compared to *Troyer citrange* rootstock. Okur et al. [11] and Anac et al. [6] suggested that this difference resulted from the nutrition properties of rootstocks. *Poncirus trifoliata* rootstock can uptake more $Na^+$ ion from soil. Can et al. [12] determined that the value of leaf area index was 1.3 in Satsuma budded onto *Poncirus trifoliata* rootstock and 1.8 in Satsuma budded onto *Troyer citrange* rootstock. When the salination rose to 6.5 dS $m^{-1}$, these values were 0.7 and 1.0 for *Poncirus trifoliata* and *Troyer citrange*, respectively. In this case, *Troyer citrange* rootstock, having more leaf area, will uptake more water from the soil and cause the moving of saline waters towards the rooting zone.

In the soil samples taken in August, it was determined that the negative relationships at 1 % significance level among the distance from stem and total soluble salt content, $EC_e$ values, the concentrations of $Na^+$ and $Cl^-$.

## Conclusion

During the first period, the total soluble salts, $EC_e$, $Na_e^+$ and $Cl_e^-$ values for both rootstocks decreased as the distance of the stem gradually increased. In addition, these parameters also decreased depending on the increased depth. Salt accumulations can occur at the soil surface (first distance and first depth), because of upward water flow induced by capillary action and evaporation from the soil surface in the first period. In the soils of the second period (October), soil salinity increased as the distance of drippers and the soil depth increased. The characteristic leaching area occurred just under the drippers caused by the moving of salts from this region and accumulating in the outside of the leaching area throughout laterals. In addition to this, it can be said that an average of 63 mm of rain fell in the second sampling period from October and leached the salts to deeper zones and carried throughout the laterals. Accumulation of salinity at the bottom of the surface (second depth) occurs as the salts are left behind by water uptake and never washed out.

## References

1 Mc Kersie BD, Leshem YY (1994) *Salt stress, stress and stress coping in cultivated plants*. Kluver Academic Publishers. The Netherlands
2 Kelley WP (1951) *Alkali soils their formation, properties and reclamation*. Reinhold Publishing Corporation, New York, USA
3 Neumann G, Pierson WJ (1966) *Principles of physical oceanography*. Prentice-Hall, Englewood Cliffs, NJ, USA
4 Jones BF, Vengosh A, Rosenthal E, Yechieli Y (1999) Geochemical investigation. In: Bear J, Cheng AH-D, Sorek S, Quazar D, Herrera I (eds) In: *Seawater intrusion in coastal aquifers-concepts, methods and practices*. Kluwer Academic Publishers, The Netherlands, 14
5 Richter BC, Kreitler CW (1993) *Geochemical techniques for identifying sources of ground water salinization*. CRC Press, Boca Raton, USA
6 Anaç D, Anaç S, Aksoy U, Okur B, Ul MA, Hepaksoy S, Kılıç CC, Dorsan F, Can Z (1999) Effect of potassium fertilization on soil salinization in Satsuma mandarin orchards. In: AE Johnston, W Maibaum (eds) In: *Balanced fertilization and crop response to potassium*. Proceedings of the Int. Sym. Of the Soil and Water Research Institute held at Tehran, Iran, 15–18 May 1999
7 Black CA (1965) *Methods of soil analysis part-II*. Amer Soc of Agronomy-Inc, Madison, Wisconsin, USA
8 Bouyoucos GJ (1962) A recalibration of the hidrometer method for making mechanical analysis of the soils. *Agronomy J* 419: 434
9 Jackson ML (1967) *Soil chemical analysis*. Prentice-Hall, New Delhi
10 Schlichting E, Blume HP (1966) *Bodenkundliches Praktikum*. Verlag Paul Paney, Hamburg und Berlin, 121–125
11 Okur B, Kılıç CC, Anaç D, Anaç S, Ul MA, Dorsan F, Aksoy U, Can HZ, Hepaksoy S (1999) Salinization effects on soil and leaf properties of Satsuma orchards. Dahlia Greidinger International Symposium. *Nutrient management under salinity and water stress*. 1–4 March 1999. Haifa, Israel
12 Can HZ, Anaç D, Kukul Y, Hepaksoy S (2002) Alleviation of salinity due to sea water intrusion by using potassium fertilization in Satsuma mandarin trees budded on two different rootstocks. $26^{th}$ International Hort. Congress. Toronto, Canada

# Index

GPSR Compliance

*The European Union's (EU) General Product Safety Regulation (GPSR) is a set of rules that requires consumer products to be safe and our obligations to ensure this.*

*If you have any concerns about our products, you can contact us on ProductSafety@springernature.com*

In case Publisher is established outside the EU, the EU authorized representative is:

Springer Nature Customer Service Center GmbH
Europaplatz 3
69115 Heidelberg, Germany

**Batch number: 10731598**

Printed by Printforce, the Netherlands